ROBOT ANALYSIS
AND CONTROL

ROBOT ANALYSIS AND CONTROL

H. ASADA and J.-J. E. SLOTINE
Massachusetts Institute of Technology

A Wiley-Interscience Publication

JOHN WILEY AND SONS
New York   Chichester   Brisbane   Toronto   Singapore

Library of Congress Cataloging in Publication Data:

Asada, H. (Haruhiko)
   Robot analysis and control.

   "A Wiley Interscience publication."
   Bibliography: p.
   Includes index.
   Robotics.   I. Slotine, J.-J. E. (Jean-Jacques E.)
   I. Title.

TJ211.A79   1985   629.8'92   86-1579
ISBN 0-471-83029-1

Printed in the United States of America

10 9 8 7 6 5 4 3

To our parents and families.

PREFACE

This book is based on the first-year graduate course on robot design and control that we developed at M.I.T., and on a Summer Session for advanced researchers and faculty members initiating their own robotics course. It covers the fundamental kinematic and dynamic analysis of manipulator arms, and the key techniques for trajectory control and compliant motion control. The book is intended to provide students and researchers with a sound scientific basis on the central aspects and concepts of robot manipulation.

Throughout the book, a strong emphasis is put on the physical meaning of the concepts introduced. The material is supported with abundant examples, most of which are adapted from successful industrial practice or advanced research topics. Carefully devised conceptual diagrams are provided to develop the reader's intuition. Each chapter includes a discussion of the current research topics where the reader is referred to the latest publications. The problem sets provided at the end of the book are tutorial in nature, and are carefully devised to develop the student's talent for synthesis. Most problems are comprehensive examples inspired from effective practical applications and advanced research topics.

Chapters 1 through 5 are easily accessible to juniors and seniors with an elementary background in mechanics and linear algebra. Chapters 6 and 7 assume that the reader has had an introductory undergraduate course in control - a brushup on basic control concepts is offered in Appendix A.6.1 . The material of Sections 2.1., 3.1.1., 3.1.2, 5.2.1, and Appendix A.6.1 is classical and may be skipped by advanced readers. On the other hand, undergraduates may only skim Sections 2.3.3, 3.2.2, 4.2.3, 5.3, 6.3.3, 6.4.2, and 7.4, and may skip Appendix A.6.2 .

This book benefited from numerous discussions with our colleagues and students. Ken Salisbury deserves special thanks for his careful reading of large parts of the manuscript and his valuable comments. Our students Neil Goldfine and especially Harry West had an active participation in the proof-reading. Bob Brodersen, Jenet McIver, Carol Beasley, Bruce Williams, and Vicki Rogers helped smooth out the production of the final text.

Haruhiko Asada and Jean-Jacques E. Slotine

Cambridge, November 1985

TABLE OF CONTENTS

ROBOT ANALYSIS
AND CONTROL

Chapter 1
INTRODUCTION

1.1. A Sense of History

Many definitions have been suggested for what we call a *robot*. The word may conjure up various levels of technological sophistication, ranging from a simple material handling device to the advanced anthropomorphic machine of science fiction. The image of robots varies widely with researchers, engineers, robot manufacturers, and countries. However, it is widely accepted that contemporary industrial robots originated in the invention of a programmed material handling device by George C. Devol. In 1954, Devol filed a U.S. patent for a new machine for part transfer, and he claimed the basic concept of *teach-in/playback* to control the device. This scheme is now extensively used in most of today's industrial robots.

Devol's industrial robots have their origins in two preceding technologies : *numerical control* for machine tools, and *remote manipulation*. Numerical control is a scheme to generate control actions based on stored data. Stored data may include coordinate data of points to which the machine is to be moved, clock signals to start and stop operations, and logical statements for branching control sequences. The whole sequence of operations and its variations are prescribed and stored in a form of memory, so that different tasks can be performed without requiring major hardware changes. Modern manufacturing systems must produce a variety of products in small batches, rather than a large number of the same products for an extended period of time, and frequent changes of product models and production schedules require *flexibility* in the manufacturing system. The transfer line approach, which is most effective for mass production, is not appropriate when such flexibility is needed (Figure 1-1). When a major product change is required, a special-purpose production line becomes useless and often ends up being abandoned, despite the large capital investment it originally involved. Flexible automation has been a central issue in manufacturing innovation

1

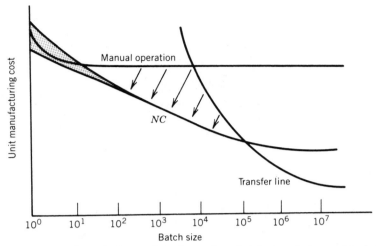

Figure 1-1 :   The general trend of manufacturing cost vs. batch size: Labor cost is dominant in manual operations using general-purpose machines, while capital investment cost is a major part of the total manufacturing cost in transfer line automation. Numerical control reduces the manufacturing cost for small-to-medium batch sizes.

for a few decades, and numerical control has played a central role in increasing system flexibility. Contemporary industrial robots are programmable machines that can perform different operations by simply modifying stored data, a feature that has evolved from the application of numerical control.

Another origin of today's industrial robots can be found in remote manipulators. A remote manipulator is a device that performs a task at a distance. It can be used in environments that human workers cannot easily or safely access, e.g. for handling radio-active materials, or in some deep sea and space applications. The first *master-slave manipulator* system was developed by 1948. The concept involves an electrically powered mechanical arm installed at the operation site, and a control joystick of geometry similar to that of the mechanical arm (Figure 1-2). The joystick has position transducers at individual joints that measure the motion of the human operator as he moves the tip of the joystick. Thus the operator's motion is transformed into electrical signals, which are transmitted to the mechanical arm and cause the same motion as the one that the human operator performed. The joystick that the operator handles is called the *master* manipulator, while the mechanical

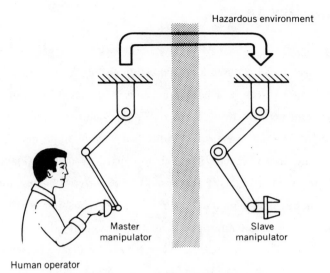

Figure 1-2 :   Master-slave manipulator.

arm is called the *slave* manipulator, since its motion is ideally the replica of the operator's commanded motion.   A master-slave manipulator has typically six degrees of freedom to allow the gripper to locate an object at an arbitrary position and orientation.   Most joints are revolute, and the whole mechanical construction is similar to that of the human arm.   This analogy with the human arm results from the need of replicating human motions.   Further, this structure allows dexterous motions in a wide range of work space, which is desirable for operations in modern manufacturing systems.

Contemporary industrial robots retain some similarity in geometry with both the human arm and remote manipulators.   Further, their basic concepts have evolved from those of numerical control and remote manipulation.   Thus a widely accepted definition of today's industrial robot is that of a numerically controlled manipulator, where the human operator and the master manipulator of Figure 1-2 are replaced by a numerical controller.

The merging of numerical control and remote manipulation creates a new field of engineering which can be referred to as *robotics*, and with it a number of scientific issues in design and control which are substantially different from those of the original technologies.

1.2. A Sense of Design

Robots are required to have much higher *mobility and dexterity* than traditional machine tools. They must be able to work in a large reachable range, access crowded places, handle a variety of workpieces, and perform flexible tasks. The unique mechanical structure of robots, which parallels that of the human arm, results from these high mobility and dexterity requirements. This structure, however, significantly departs from traditional machine design. A robot mechanical structure is basically composed of cantilevered beams, forming a sequence of arm links connected by hinged joints. Such a structure has inherently poor mechanical stiffness and accuracy, hence is not appropriate for the heavy-duty, high-precision applications required of machine tools. Further, it also implies a serial sequence of servoed joints, whose errors accumulate along the linkage. In order to exploit the high mobility and dexterity uniquely featured by the serial linkage, these difficulties must be overcome by advanced design and control techniques.

The serial linkage geometry of manipulator arms is described by complex nonlinear equations. Effective analytical tools are necessary to understand the geometric and kinematic behavior of the manipulator, globally referred to as the manipulator *kinematics*. This represents an important and unique area of robotis research, since research in kinematics and design has traditionally focused upon single-input mechanisms with single actuators moving at constant speeds, while robots are multi-input spatial mechanisms which require more sophisticated analytical tools.

The *dynamic* behavior of robot manipulators is also complex, since the dynamics of multi-input spatial linkages are highly coupled and nonlinear. The motion of each joint is significantly affected by the motions of all the other joints. The inertial load imposed to each joint varies widely depending on the configuration of the manipulator arm. Coriolis and centrifugal effects are prominent when the manipulator arm moves at high speeds. The kinematic and dynamic complexities create unique control problems that are not adequately handled by standard linear control techniques, and thus make effective *control system design* a critical issue in robotics.

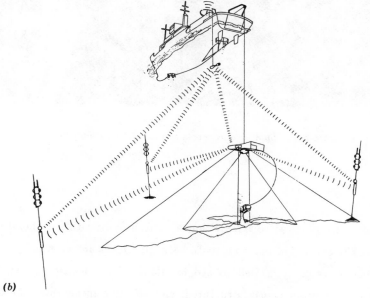

Figure 1-3 : (a) The Remote Manipulator System on board the NASA Space Shuttle, as viewed by its operator (*Courtesy of NASA*); (b) The ARGO/JASON project of Woods Hole Oceanographic Institution, to be completed about 1988 — the operator on the ground commands in real-time an underwater manipulator deeply submerged at 6000 meters under sea level (*Courtesy of W.H.O.I., Woods Hole, Mass.*).

Figure 1-4 :   Spot welding robots in a car body assembly line.

Finally, robots are required to *interact* much more heavily with peripheral devices than traditional numerically-controlled machine tools. Machine tools are essentially self-contained systems that handle workpieces in well-defined locations. By contrast, the environment in which robots are used is often poorly structured, and effective means must be developed to identify the locations of the workpieces as well as to communicate to peripheral devices and other machines in a coordinated fashion.

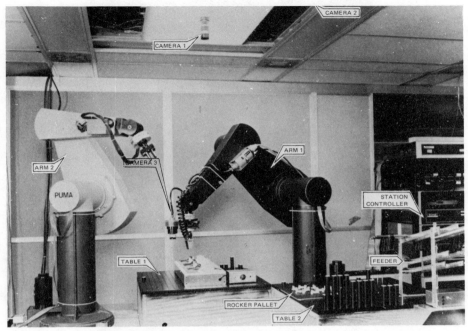

Figure 1-5 : Vision-guided assembly robots. Two video cameras are installed on the ceiling, while a third camera is carried by Arm 1 (S.R.I.).

Robots are also critically different from master-slave manipulators, in that they are *autonomous* systems. Master-slave manipulators are essentially manually controlled systems, where the human operator takes the decisions and applies control actions. The operator interprets a given task, finds an appropriate strategy to accomplish the task, and plans the procedure of operations. He devises an effective way of achieving the goal on the basis of his experience and knowledge about the task. His decisions are then transferred to the slave manipulator through the joystick. The resultant motion of the slave manipulator is monitored by the operator, and necessary adjustments or modifications of control actions are provided when the resultant motion is not adequate, or when unexpected events occur during the operation. The human operator is, therefore, an essential part of the control loop. When the operator is eliminated from the control system, all the planning and control commands must be generated by the machine itself. The detailed procedure of operations must be set up in advance, and each step of motion command must be generated and coded in an appropriate

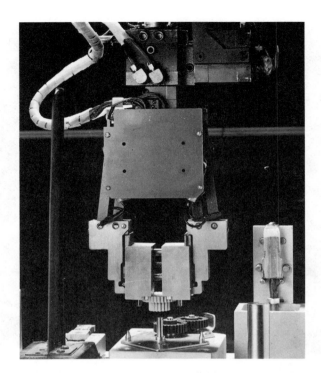

Figure 1-6 : Gear plate assembly. The robot uses tactile feedback to mesh gears .

form so that the robot can interpret it and execute it accurately. Effective means to store the commands and manage the data file are also needed . Thus, *programming and command generation* are critical issues in robotics. In addition, the robot must be able to fully monitor its own motion. In order to adapt to disturbances and unpredictable changes in the work environment, the robot needs a variety of *sensors,* so as to obtain information both about the environment (using *external* sensors, such as cameras or touch sensors) and about itself (using *internal* sensors, such as joint encoders or joint torque sensors). Effective sensor-based strategies that incorporate this information require advanced control algorithms. But they also imply a detailed understanding of the task.

1.3. Manipulators and Manipulation

Contemporary industrial needs drive the applications of robots to ever more advanced tasks. Robots are required to perform highly skilled jobs with minimum human assistance or

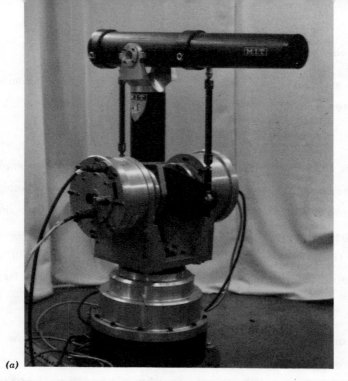

(a)

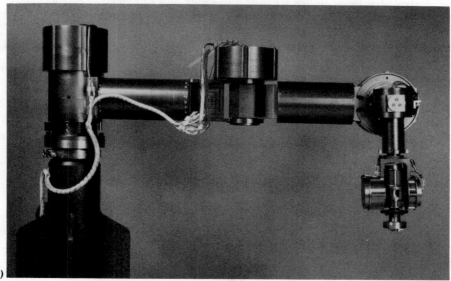

(b)

Figure 1-7: The M.I.T. Direct-Drive Arm III [a], and the C.M.U. Direct-Drive Arm II (*Courtesy of T. Kanade, Carnegie-Mellon University*) [b]. Direct-drive is an innovative manipulator design where the arm links are directly coupled to high-torques motors, without gear reducers. The absence of gearing eliminates backlash problems, and yields high mechanical stiffness and low friction. The arms shown achieve high speeds of over 10 m/s, and 5 *G* accelerations.

intervention. To extend the applications and abilities of robots, it becomes important to develop a sound understanding of the *tasks* themselves.

In order to devise appropriate arm mechanisms and to develop effective control algorithms, we need to precisely understand how a given task should be accomplished and what sort of motions the robot should be able to achieve. To perform an assembly operation, for example, we need to know how to guide the assembly part to the desired location, mate it with another part, and secure it in an appropriate way. In a grinding operation, the robot must properly position the grinding wheel while accommodating the contact force. We need to analyze the grinding process itself in order to generate appropriate force and motion commands.

A detailed understanding of the underlying principles and "know-how" involved in the task must be developed in order to use industrial robots effectively, while there is no such need for making control strategies *explicit* when the assembly and grinding operations are performed by a human worker. Human beings perform sophisticated manipulation tasks without being aware of the control principles involved. We have *trained* ourselves to be capable of skilled jobs, but in general we do not know what the acquired skills are exactly about. A sound and

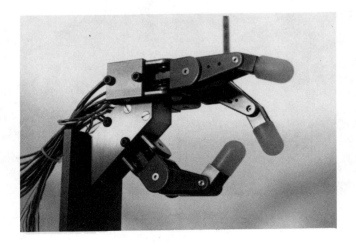

Figure 1-8 : Nine degree-of-freedom robot hand. This articulated hand is being used to study control, sensing, and language issues at the M.I.T. Artificial Intelligence Laboratory. *Photograph courtesy of D. R. Lampe, The MIT Report.*

explicit understanding of manipulation operations, however, is essential for the long-term progress of robotics. This scientific aspect of manipulation has never been studied systematically before, and represents an emerging and important part of robotics research.

The goal of robotics is thus two-fold: to extend our understanding about manipulation and tasks, and to develop engineering methodologies to actually perform the desired manipulation tasks. The central aspect of the latter goal is to devise design and control techniques for manipulator arms. This is the focus of this book.

1.4. Robot Analysis and Control in a Nutshell

This section briefly describes the contents of each chapter of this book. In addition to its economic importance, robotics has a "classical" scientific flavor that makes it a particularly rewarding field of study.

Consider the two-link planar manipulator of Figure 1-9, which can be regarded as the simplest nontrivial manipulator, and assume that we want to use the robot to position some object located in its end-effector. In order to do so, we must first find joint angles θ_1 and θ_2 that make the manipulator endpoint coincide with the desired location. This problem is referred to as the *inverse kinematics* problem. For the manipulator of Figure 1-9, there are two *configurations* (i.e., two sets of joint angles θ_1 and θ_2) that lead to the same endpoint position,

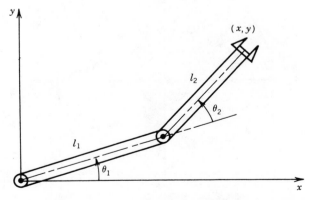

Figure 1-9 : A two-link articulated planar manipulator. Desired end-effector positions are reached by moving the two motors located at the joints.

as illustrated in Figure 1-10. Note that of course not all points in the plane are reachable by the manipulator — the set of reachable points is called the manipulator *workspace*, and is shown in Figure 1-11. Manipulator kinematics, discussed in Chapters 2 and 3, is a study of such *geometric* problems.

When the manipulator is at rest but in contact with its environment (say a car body), it may also be required to apply certain static *forces* at its end-effector (Figure 1-12). This problem is discussed in Chapter 4.

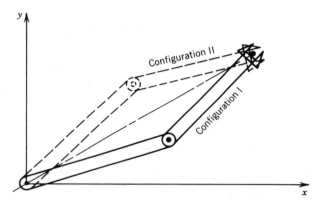

Figure 1-10 : The two inverse kinematics solutions.

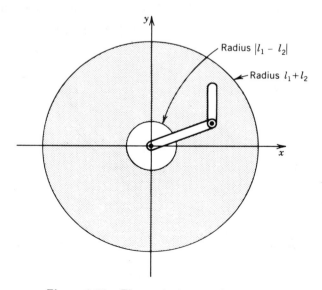

Figure 1-11 : The manipulator workspace.

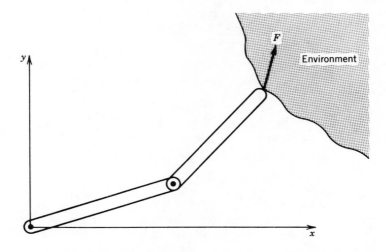

Figure 1-12 : Exerting a force on the environment.

Once we know from our kinematic study *where* to move the manipulator joints in order to attain a desired position, we must find out *how* to actually get there, in other words, analyze the manipulator *dynamics* (Chapter 5) so as to design an adequate *control system* (Chapter 6). The problem may become fairly sophisticated if the manipulator is required to accurately follow a desired fast *trajectory* (as in plasma welding or laser cutting, for example), as opposed to simply going to a fixed position (as in pick-and-place or spot-welding operations, for instance).

Finally, the problem of controlling both the manipulator motion and its force interactions with the environment is discussed in Chapter 7. A complete understanding of this problem, referred to as *compliant motion control*, is really at the core of robot manipulation, and should allow robots to replace humans in increasingly complex assembly and manipulation tasks.

Chapter 2
KINEMATICS I: GEOMETRY

Manipulator kinematics is a study of the geometry of manipulator arm motions. Since the performance of specific tasks is achieved through the movement of the manipulator arm linkages, kinematics is a fundamental tool in manipulator design and control. In this chapter, the mathematical tools required to describe arm linkage motion are developed. Also, the fundamental equations that govern kinematic behavior are derived and the solution of these equations is discussed.

2.1. Mathematical Preliminary

2.1.1. Position and Orientation of a Rigid Body

The arm linkage of a manipulator can be modeled as a system of rigid bodies. The location of each single rigid body is completely described by its *position* and *orientation*.

The position can be represented by the coordinates of an arbitrary point fixed with respect to the rigid body. Let $O-xyz$ be a coordinate frame fixed to the ground and let point O' be an arbitrary point fixed to the rigid body, as shown in Figure 2-1. Then the position of the rigid body is represented with reference to the coordinate frame $O-xyz$ by

$$\mathbf{x}_0 = \begin{pmatrix} x_0 \\ y_0 \\ z_0 \end{pmatrix} \tag{2-1}$$

where $\mathbf{x}_0$ is a 3×1 column vector.

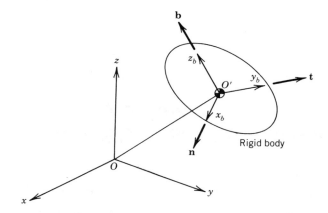

Figure 2-1 : Position and orientation of a rigid body.

To represent the orientation of the rigid body, three coordinate axes x_b, y_b, and z_b are attached to the rigid body as shown in the figure. These axes form another coordinate frame $O'-x_b y_b z_b$, which moves with the rigid body. The orientation of the rigid body is then represented by the directions of these coordinate axes. Let **n**, **t** and **b** be unit vectors pointing the directions of the coordinate axes, x_b, y_b and z_b, respectively. The components of each unit vector are direction cosines of each coordinate axis projected into the fixed coordinate frame $O-xyz$. For convenience, we combine the three vectors together and write them using the 3×3 matrix **R**:

$$\mathbf{R} = [\, \mathbf{n, t, b} \,] \tag{2-2}$$

The matrix **R** completely describes the orientation of the rigid body with reference to the fixed coordinate frame $O-xyz$. Note that the column vectors of matrix **R** are orthogonal to each other

$$\mathbf{n}^T \mathbf{t} = 0 \qquad\qquad \mathbf{t}^T \mathbf{b} = 0 \qquad\qquad \mathbf{b}^T \mathbf{n} = 0 \tag{2-3}$$

and further have unit length

$$|\mathbf{n}| = 1 \qquad\qquad |\mathbf{t}| = 1 \qquad\qquad |\mathbf{b}| = 1 \tag{2-4}$$

(where $|\mathbf{a}|$ designates the Euclidian norm of a vector **a**). Such a matrix, in which all the

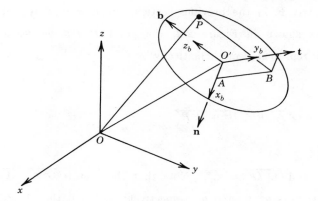

Figure 2-2 : Coordinate transformation.

column vectors have unit length and are orthogonal to each other, is referred to as an *orthonormal matrix.*

2.1.2. Coordinate Transformations

Let P be an arbitrary point in space, as shown in Figure 2-2. We represent the coordinates of point P with reference to the fixed frame $0-xyz$ by

$$\mathbf{x} = \begin{pmatrix} x \\ y \\ z \end{pmatrix} \qquad (2-5)$$

The position of point P can be also represented with reference to the coordinate frame fixed to the rigid body, $O'-x_b y_b z_b$, by

$$\mathbf{x}^b = \begin{pmatrix} u \\ v \\ w \end{pmatrix} \qquad (2-6)$$

The superscript ^{b} indicates that the vector is defined with reference to the body coordinate frame. Let us now find the relationship between the two coordinate systems. This relationship defines the coordinate transformation between the fixed frame and the body coordinate frame. The position and orientation of the rigid body, which are represented by the 3×1 vector $\mathbf{x}_0$

and the 3 × 3 matrix $\mathbf{R}$ in the previous section, are now used to obtain the coordinate transformation.   As shown in Figure 2-2, the point P can be reached through points O', A and B. This is represented mathematically by

$$\overrightarrow{O\,P} = \overrightarrow{O\,O'} + \overrightarrow{O'A} + \overrightarrow{A\,B} + \overrightarrow{B\,P} \qquad (2-7)$$

where $\overrightarrow{O\,P} = \mathbf{x}$ and $\overrightarrow{O\,O'} = \mathbf{x}_0$. Note that the vectors $\overrightarrow{O'A}$, $\overrightarrow{A\,B}$ and $\overrightarrow{B\,P}$ are parallel to the unit vectors $\mathbf{n}$, $\mathbf{t}$, and $\mathbf{b}$, respectively, and that their lengths are given by u , v , and w. Thus, we can rewrite the above expression as:

$$\mathbf{x} = \mathbf{x}_0 + u\mathbf{n} + v\mathbf{t} + w\mathbf{b} \qquad (2-8)$$

i.e., from (2-2) and (2-6), as

$$\mathbf{x} = \mathbf{x}_0 + \mathbf{R}\mathbf{x}^b \qquad (2-9)$$

Equation (2-9) provides the desired coordinate transformation from the body coordinates $\mathbf{x}^b$ to the fixed coordinates $\mathbf{x}$. Note that this coordinate transformation is given in terms of $\mathbf{x}_0$ and $\mathbf{R}$, which represent the position and orientation of the rigid body, or of the body coordinate frame relative to the fixed frame.

Let us premultiply both sides of equation (2-9) by the transpose $\mathbf{R}^T$ of matrix $\mathbf{R}$,

$$\mathbf{R}^T\mathbf{x} = \mathbf{R}^T\mathbf{x}_0 + \mathbf{R}^T\mathbf{R}\mathbf{x}^b \qquad (2-10)$$

From (2-3) and (2-4), the matrix product $\mathbf{R}^T\mathbf{R}$ on the right-hand side becomes

$$\mathbf{R}^T\mathbf{R} = \begin{bmatrix} \mathbf{n}^T\mathbf{n} & \mathbf{n}^T\mathbf{t} & \mathbf{n}^T\mathbf{b} \\ \mathbf{t}^T\mathbf{n} & \mathbf{t}^T\mathbf{t} & \mathbf{t}^T\mathbf{b} \\ \mathbf{b}^T\mathbf{n} & \mathbf{b}^T\mathbf{t} & \mathbf{b}^T\mathbf{b} \end{bmatrix} = \begin{bmatrix} 1 & 0 & 0 \\ 0 & 1 & 0 \\ 0 & 0 & 1 \end{bmatrix} \qquad (2-11)$$

Therefore, equation (2-10) reduces to

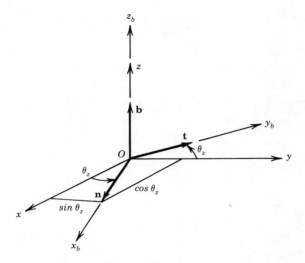

Figure 2-3 : Example 2-1.

$$\mathbf{x}^b = -\mathbf{R}^T\mathbf{x}_0 + \mathbf{R}^T\mathbf{x} \tag{2-12}$$

Equation (2-12) represents the coordinate transformation from the fixed coordinates to the body coordinates, that is, the inverse of the original transformation (2-9). Thus, the inverse transformation is simply obtained by using the transpose of the matrix $\mathbf{R}$. Also, as equation (2-11) shows, the inverse of an orthonormal matrix is simply given by the transposed matrix:

$$\mathbf{R}^{-1} = \mathbf{R}^T \tag{2-13}$$

Example 2-1

As shown in Figure 2-3, the origin of coordinate frame $O'-x_b y_b z_b$ coincides with the origin of the fixed frame $O-xyz$. The angle between axes x and x_b is denoted by $\theta_z = \angle xOx_b$. Axis z_b , on the other hand, coincides with axis z. Let us find the vector $\mathbf{x}_0$ and the matrix $\mathbf{R}$ that represent the position and orientation of frame $O'-x_b y_b z_b$ relative to frame $O-xyz$, and then obtain the coordinate transformation from $O'-x_b y_b z_b$ to $O-xyz$.

Since the origins of the two coordinate frames coincide, position vector $\mathbf{x}_0$ is zero. To obtain the rotation matrix $\mathbf{R}$, let us find the three unit vectors, $\mathbf{n}$, $\mathbf{t}$, and $\mathbf{b}$, composing $\mathbf{R}$. As shown in Figure 2-3, the components of each vector are its direction cosines with respect to $O-xyz$. Therefore,

$$\mathbf{n} = \begin{pmatrix} \cos\theta_z \\ \sin\theta_z \\ 0 \end{pmatrix} \qquad \mathbf{t} = \begin{pmatrix} -\sin\theta_z \\ \cos\theta_z \\ 0 \end{pmatrix} \qquad \mathbf{b} = \begin{pmatrix} 0 \\ 0 \\ 1 \end{pmatrix}$$

so that

$$\mathbf{R} = \begin{bmatrix} \cos\theta_z & -\sin\theta_z & 0 \\ \sin\theta_z & \cos\theta_z & 0 \\ 0 & 0 & 1 \end{bmatrix} \qquad (2-14)$$

The coordinate transformation is then obtained by substituting the matrix $\mathbf{R}$ and $\mathbf{x}_0 = 0$ into equation (2-9). The components of the transformation expressions are thus given by

$$x = u\cos\theta_z - v\sin\theta_z$$

$$y = u\sin\theta_z + v\cos\theta_z \qquad (2-15)$$

$$z = w$$

Let us verify the above results. Figure 2-4 shows a two-dimensional view of the two coordinate frames. The point P' is the projection of point P onto the xy plane. Points A and B are projections of the point P' onto axes x and x_b, respectively, and point C is the projection of point B onto the x axis. From this figure, the above equations of the coordinate transformation can be interpreted as follows. We have

$$x = \overline{OA} = \overline{OC} - \overline{AC}$$

$$= \overline{OB}\cos\angle BOC - \overline{P'B}\sin\angle AP'B$$

$$= u\cos\theta_z - v\sin\theta_z$$

which agrees with the first equation of (2-15). The other equations can be derived in the same way.   ▲▲▲

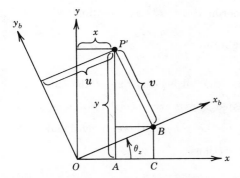

Figure 2-4 : 'Two-dimensional coordinate transformation.

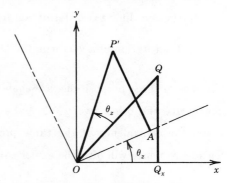

Figure 2-5 :   Rotation of Point P'.

Example 2-2

Figure 2-5 shows the same two-dimensional coordinate frames and arbitrary point P' as in the previous figure. Let Q be the point in the x-y plane whose coordinates with reference to $O-xy$ are the same as the coordinates of point P' with respect to $O-x_b y_b$. Namely, $\overline{OQ}_x = \overline{OA}$ and $\overline{QQ}_x = \overline{P'A}$, as shown in Figure 2-5. The problem is to show that point P' is reached by rotating point Q about the origin O by an angle θ_z.

From the figure, $\overline{OP'} = \overline{OQ}$ and $\angle P'OA = \angle QOQ_x$. Therefore, $\angle QOP' = \angle Q_x OA = \theta_z$, and point P' is obtained by the rotation θ_z of point Q about the origin O. This discussion yields another interpretation of equation (2-15) and matrix $\mathbf{R}$. If we regard u and v in (2-15) as the coordinates of point Q in $O-xy$, then (2-15) provides the coordinates of the point P' obtained by rotating point Q about the origin by an angle θ_z. In the three-dimensional space shown in Figure 2-3, this rotation is about the z axis. Consequently, the matrix $\mathbf{R}$ associated with equation (2-15) represents the rotation about the z-axis, and therefore is called the *rotation matrix*.   ▵▵▵

In summary, the rotation matrix $\mathbf{R}$ has three distinct physical meanings. It can represent

(1) the orientation of the coordinate frame $O'-x_b y_b z_b$ relative to $O-xyz$, where the column vector represents the direction cosines of each axis of $O'-x_b y_b z_b$ projected into the $O-xyz$ frame,

(2) the coordinate transformation from $O'-x_b y_b z_b$ coordinates to $O-xyz$ coordinates,

(3) the rotation of a vector in the $O-xyz$ coordinate frame.

The above three propositions are equivalent in the sense that, using any one of the three, we can derive the other two. While the examples discussed above are special two-dimensional cases, the equivalence of the three propositions holds in the general three-dimensional case. This equivalence will often be exploited in the following sections.

2.1.3. Euler Angles

In the previous sections, we used the 3×3 matrix $\mathbf{R}$ to represent the orientation of a rigid body or a coordinate frame attached to the body. The elements of the matrix, however, are not independent. The matrix has nine elements in total, which are all subject to the orthogonality conditions (2-3) and the unit length conditions (2-4). Since there are six of these conditions, only three of the nine elements are independent. In other words, the matrix representation of orientation is *redundant*. In this section, a representation of the rigid body orientation that only uses three independent variables is discussed.

Consider the three rotations of frame $O-xyz$ shown in Figure 2-6. First, the coordinate frame is rotated about the z axis by an angle ϕ (Figure 2-6.a). Secondly, the new coordinate frame $O-x'y'z$ is rotated about the x' axis by an angle θ (Figure 2-6.b). Finally, the newest frame $O-x'y''z''$ is then rotated about the z'' axis by an angle ψ. The resultant coordinate frame $O-x_b y_b z_b$ is shown in Figure 2-6.c . The three angles, ϕ, θ and ψ, determine the orientation of the coordinate frame uniquely, and are referred to as *Euler angles*. The Euler angles are independent, in that each of them can vary arbitrarily.

For a given arbitrary orientation of coordinate frame $O-x_b y_b z_b$, the Euler angles can be determined as follows. Let line Ox' in Figure (c) be the intersection between the x_b-y_b plane and the $x-y$ plane. This intersection is referred to as the *line of nodes*. The angle ψ is defined as the angle from the line of nodes to the x_b axis in Figure (c), while the angle ϕ is the angle from the x axis to the line of nodes. The angle θ, on the other hand, is defined as the angle

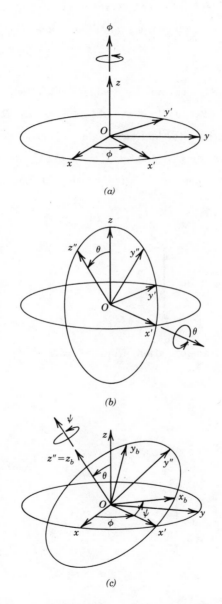

Figure 2-6 : The three consecutive rotations used to define the Euler angles.

from the z axis to the $z_b = z''$ axis. All the angles are measured in a right-hand sense. Thus the three angles can be defined for an arbitrary orientation of coordinate frame $O-x_b y_b z_b$ relative to the fixed frame $O-xyz$. Note, however, that the Euler angles are not unique. The set of angles $(\phi + \pi, -\theta, \psi + \pi)$ gives the same orientation as (ϕ, θ, ψ), as later discussed in Section 2.3.2.

The Euler angles are independent variables which determine the orientation of a coordinate frame uniquely. Let us find the rotation matrix $\mathbf{R}$ that represents the three consecutive rotations associated with the Euler angles. Consider the coordinate transformation associated with the rotation ϕ. Coordinates $\mathbf{x'} = [x',y',z]^T$ are transformed to coordinates $\mathbf{x} = [x,y,z]^T$ by the 3×3 rotation matrix $\mathbf{R}_z(\phi)$, which is defined as

$$\mathbf{x} = \mathbf{R}_z(\phi)\mathbf{x'}$$

$$\mathbf{R}_z(\phi) = \begin{bmatrix} \cos\phi & -\sin\phi & 0 \\ \sin\phi & \cos\phi & 0 \\ 0 & 0 & 1 \end{bmatrix} \tag{2-16}$$

Similarly, the coordinate transformation from $\mathbf{x''} = [x',y'',z'']^T$ to $\mathbf{x'}$ associated with the rotation θ is given by

$$\mathbf{x'} = \mathbf{R}_{x'}(\theta)\mathbf{x''}$$

where

$$\mathbf{R}_{x'}(\theta) = \begin{bmatrix} 1 & 0 & 0 \\ 0 & \cos\theta & -\sin\theta \\ 0 & \sin\theta & \cos\theta \end{bmatrix} \tag{2-17}$$

Finally, for the rotation ψ, we have

$$\mathbf{x''} = \mathbf{R}_{z''}(\psi)\mathbf{x}^b$$

$$\mathbf{R}_{z''}(\psi) = \begin{bmatrix} \cos\psi & -\sin\psi & 0 \\ \sin\psi & \cos\psi & 0 \\ 0 & 0 & 1 \end{bmatrix} \tag{2-18}$$

Combining the three coordinate transformations yields

$$\mathbf{x} = \mathbf{R}_z(\phi)\,\mathbf{R}_{x'}(\theta)\,\mathbf{R}_{z''}(\psi)\,\mathbf{x}^b \tag{2-19}$$

Let us replace the above matrix product by

$$\mathbf{R}(\phi,\theta,\psi) = \mathbf{R}_z(\phi)\,\mathbf{R}_{x'}(\theta)\,\mathbf{R}_{z''}(\psi) \tag{2-20}$$

The matrix $\mathbf{R}(\phi, \theta, \psi)$ provides the coordinate transformation from $\mathbf{x}^b$ to $\mathbf{x}$. As a result of the equivalency between the coordinate transformation matrix and the rotation matrix discussed in Section 2.1.2, the matrix $\mathbf{R}(\phi, \theta, \psi)$ represents the rotation from coordinate frame $O-xyz$ to coordinate frame $O-x_b y_b z_b$. Also, the column vectors of $\mathbf{R}(\phi, \theta, \psi)$ represent the direction cosines of the coordinate axes x_b, y_b and z_b with reference to the $O-xyz$ frame.

Another independent set of angles, consisting of *roll, pitch,* and *yaw,* is widely used in robotics to describe rigid body orientation. The roll angle θ_z represents a rotation about a z axis, while the pitch and yaw angles represent consecutive rotations about corresponding y and x axes. Using the notations of Figure 2-6, the rotation matrix associated to roll-pitch-yaw is $\mathbf{R}_z(\theta_z)\,\mathbf{R}_{y'}(\theta_y)\,\mathbf{R}_{x''}(\theta_x)$.

2.1.4. Homogeneous Transformations

In this section, we develop a useful method for representing coordinate transformations in a compact form.

Let us recall the coordinate transformation given by equation (2-9):

$$\mathbf{x} = \mathbf{x}_0 + \mathbf{R}\mathbf{x}^b \tag{2-21}$$

The first term of the right-hand side represents the translational transformation, while the second term represents the rotational transformation. The goal of this section is to derive a simple representation of the coordinate transformation in which both the translational and rotational transformations are given by a single matrix. To this end, let us define the 4×1 vectors:

$$\mathbf{X} = \begin{bmatrix} x \\ y \\ z \\ 1 \end{bmatrix} \qquad \mathbf{X}^b = \begin{bmatrix} u \\ v \\ w \\ 1 \end{bmatrix}$$

and the 4×4 matrix:

$$\mathbf{A} = \left[\begin{array}{c|c} \mathbf{R} & \mathbf{x}_0 \\ \hline \mathbf{O} & 1 \end{array} \right] \tag{2--22}$$

The original vectors $\mathbf{x}$ and $\mathbf{x}^b$ are augmented by adding a "1" as the fourth element so that the result is a 4×1 vector. Also, the rotation matrix $\mathbf{R}$ is extended to a 4×4 matrix by combining it with the 3×1 position vector $\mathbf{x}_0$, with three 0's and a 1 in the fourth row. Equation (2-21) can then be written as

$$\mathbf{X} = \mathbf{A} \, \mathbf{X}^b \tag{2--23}$$

that is,

$$\begin{bmatrix} x \\ y \\ z \\ 1 \end{bmatrix} = \left[\begin{array}{c|c} \mathbf{R} & \mathbf{x}_0 \\ \hline \mathbf{O} & 1 \end{array} \right] \begin{bmatrix} u \\ v \\ w \\ 1 \end{bmatrix} \tag{2--24}$$

Note that the 4×4 matrix $\mathbf{A}$ represents both the position and orientation of the frame $O-x_b y_b z_b$. The two terms on the right-hand side of equation (2-21) are reduced to the single

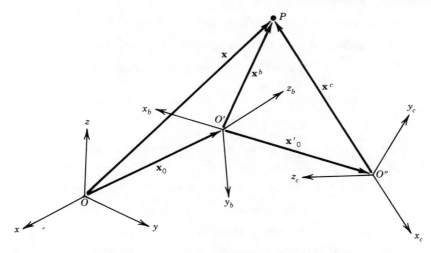

Figure 2-7 : Consecutive coordinate transformations.

term in equation (2-23). The coordinate transformation given by equation (2-23) is referred to as the *homogeneous transformation*.

The compactness of the homogeneous transformation is particularly advantageous when we represent consecutive transformations. Let $O''-x_c y_c z_c$ be another coordinate frame, as shown in Figure 2-7, and $\mathbf{x}^c$ be the coordinates of point P with reference to $O''-x_c y_c z_c$. Then

$$\mathbf{x}^b = \mathbf{x}'_0 + \mathbf{R}'\mathbf{x}^c \tag{2-25}$$

where $\mathbf{x}'_0$ and $\mathbf{R}'$ are the 3×1 vector and 3×3 matrix associated with the coordinate transformation from $\mathbf{x}^c$ to $\mathbf{x}^b$. Substituting (2-25) into (2-21), we obtain

$$\mathbf{x} = \mathbf{x}_0 + \mathbf{R}\mathbf{x}'_0 + \mathbf{R}\mathbf{R}'\mathbf{x}^c \tag{2-26}$$

There are now three terms in the right-hand side of equation (2-26). As the transformation is repeated, the number of terms in the right-hand side increases. In general, n consecutive coordinate transformations lead to a n-th order polynomial consisting of $(n+1)$ non-homogeneous terms. The homogeneous transformation which uses the 4×4 matrix, on the other hand, provides a compact form that represents any consecutive transformation with a single term. Consider n consecutive transformation from frame n back to frame 0. Let $\mathbf{A}_i^{i-1}$ be the 4×4 matrix associated with the homogeneous transformation from frame i to frame $i-1$; then a position vector $\mathbf{X}^n$ in frame n is transformed to $\mathbf{X}^0$ in frame 0 by

$$\mathbf{X}^0 = \mathbf{A}_1^0 \, \mathbf{A}_2^1 \, \cdots \, \mathbf{A}_n^{n-1} \, \mathbf{X}^n \tag{2-27}$$

Thus the consecutive transformations are compactly described by a single term.

The 4×4 matrices have two other properties equivalent to those discussed earlier for rotation matrices. A 4×4 matrix represents the position and orientation of a coordinate frame. It also represents the translation and rotation of the coordinate frame. Thus, the equivalence property for rotation matrices also holds for 4×4 matrices, in which both translations and rotations are involved.

Example 2-3

In this example, we derive an expression of the inverse of the 4×4 matrix $\mathbf{A}$, using the corresponding inverse coordinate transformation.

The inverse coordinate transformation is given from (2-23) by

$$\mathbf{X}^b = \mathbf{A}^{-1}\mathbf{X} \tag{2-28}$$

From equation (2-12), the same inverse transformation can be expressed in the 3×3 matrix form

$$\mathbf{x}^b = -\mathbf{R}^T\mathbf{x}_0 + \mathbf{R}^T\mathbf{x} \tag{2-29}$$

Let us now convert the above expression into the 4×4 matrix form, and determine the matrix $\mathbf{A}^{-1}$. Comparing equation (2-29) with equation (2-21), we find that $\mathbf{x}_0$ of (2-21) is replaced by $-\mathbf{R}^T\mathbf{x}_0$ in (2-29), while $\mathbf{R}$ is simply replaced by $\mathbf{R}^T$. Applying the same conversion as in equation (2-22), we obtain

$$\mathbf{A}^{-1} = \left[\begin{array}{c|c} \mathbf{R}^T & -\mathbf{R}^T\mathbf{x}_0 \\ \hline \mathbf{O} & 1 \end{array}\right]$$

(2−30)

The above result can also be proved by checking that both products $A^{-1}A$ and AA^{-1} are indeed equal to the identity matrix. ▵▵▵

2.2. Kinematic Modeling of Manipulator Arms

2.2.1. Open Kinematic Chains

The mathematical tools that we developed in the previous section are now applied to the kinematic modeling of manipulator arms. In particular, we use the homogeneous tranformation to describe the position and orientation of each link member involved in a manipulator arm.

A manipulator arm is basically a series of rigid bodies in a kinematic structure. Figure 2-8 shows a manipulator arm modeled as a serial linkage of rigid bodies. Such a linkage with a

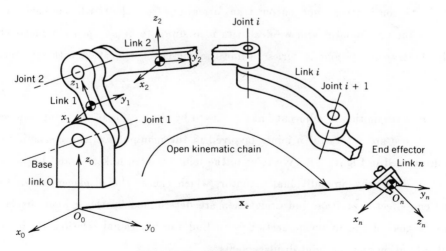

Figure 2-8 : Open kinematic chain.

serial or "open loop" structure is referred to as an *open kinematic chain*. Most of existing industrial robots and research arms are open kinematic chains, or equivalent in structure. The manipulator arms discussed in this chapter are all assumed to be open kinematic chains.

Each link member of the open kinematic chain can be numbered in series from 0 to n , as shown in Figure 2-8. The base link, which is usually fixed to the ground, is numbered 0 for convenience. The most distal link, on the other hand, is numbered n. Since a manipulator arm performs a task through the motion of an end-effector attached to the last link, our primary interest is to analyze the motion of the last link.

In order to represent the position and orientation of the end-effector, we attach a coordinate frame $O_n-x_n y_n z_n$ to the last link. The location of the coordinate frame is described with reference to another frame $O_0-x_0 y_0 z_0$, fixed to the ground (i.e. the base link), as shown in the figure. The end-effector motion is caused by motions of the intermediate links between the base link and the last link. Thus, the end-effector location can be determined by investigating the position and orientation of each link member in series from the base to the tip. To this end, we attach a coordinate frame to each of the links, namely frame $O_i-x_i y_i z_i$ to link i. We describe the position and orientation of frame $O_i-x_i y_i z_i$ relative to the previous frame $O_{i-1}-x_{i-1} y_{i-1} z_{i-1}$ by using the 4×4 matrix describing the homogeneous transformation between these frames. The end-effector position and orientation is then obtained by the consecutive homogeneous transformations from the last frame back to the base frame. Since the manipulator arm we deal with is assumed to be an open kinematic chain, we can apply the transformations in series to find the end-effector location with reference to the base frame.

The relative motion of adjacent links is caused by the motion of the joint connecting the two links. There are a total of n joints involved in the manipulator arm consisting of $(n+1)$ links, as illustrated in Figure 2-8. We refer to the joint between link i - 1 and link i as joint i. Each joint is driven by an individual actuator, which causes the displacement of the joint. Thus, the end-effector position and orientation are determined by the n joint displacements. The primary goal of the following section is to find the functional relationship between the end-effector location and the joint displacements.

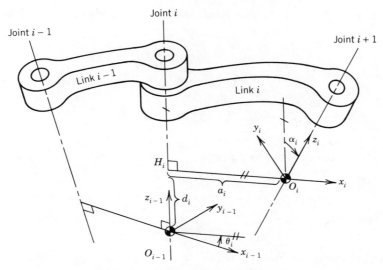

Figure 2-9 :   The Denavit-Hartenberg notation.

2.2.2. The Denavit-Hartenberg Notation

In this section we discuss the description of the kinematic relationship between a pair of adjacent links involved in an open kinematic chain.   The *Denavit-Hartenberg* notation is introduced as a systematic method of describing this kinematic relationship.   The method is based on the 4×4 matrix representation of rigid body position and orientation.   It uses a minimum number of parameters to completely describe the kinematic relationship.

Figure 2-9 shows a pair of adjacent links, link i - 1 and link i, and their associated joints, joints i - 1, i and $i+1$.   Line H_iO_i in the figure is the common normal to joint axes i and $i+1$.   The relationship between the two links is described by the relative position and orientation of the two coordinate frames attached to the two links.   In the Denavit-Hartenberg notation, the origin of the i-th coordinate frame O_i is located at the intersection of joint axis $i+1$ and the common normal between joint axes i and $i+1$, as shown in the figure.   Note that the frame of link i is at joint $i+1$ rather than at joint i.   The x_i axis is directed along the extention line of the common normal, while the z_i axis is along the joint axis $i+1$.   Finally, the y_i axis is chosen such that the resultant frame $O_i - x_i y_i z_i$ forms a right-hand coordinate system.

The relative location of the two frames can be completely determined by the following four parameters:

a_i the length of the common normal

d_i the distance between the origin O_{i-1} and the point H_i

α_i the angle between the joint axis i and the z_i axis in the right-hand sense

θ_i the angle between the x_{i-1} axis and the common normal H_iO_i measured about the z_{i-1} axis in the right-hand sense.

The parameters a_i and α_i are *constant* parameters that are determined by the geometry of the link: a_i represents the link length and α_i is the twist angle between the two joint axes. *One* of the other two parameters d_i and θ_i varies as the joint moves.

There are two types of joint mechanisms used in manipulator arms: *revolute* joints in which the adjacent links rotate with respect to each other about the joint axis, and *prismatic* joints in which the adjacent links translate linearly to each other along the joint axis. For a revolute joint, parameter θ_i is the variable that represents the joint displacement, while parameter d_i is constant. For a prismatic joint, on the other hand, parameter d_i is the variable representing the joint displacement, while θ_i is constant.

Let us now formulate the kinematic relationship between the adjacent links using 4×4 matrices. Using the equivalence property discussed in Section 2.1, the 4×4 matrix representing the location of frame i relative to frame $i\text{-}1$ can be determined by considering the associated coordinate transformation. Figure 2-10 shows the two coordinate frames, and $O_i-x_iy_iz_i$ and $O_{i-1}-x_{i-1}y_{i-1}z_{i-1}$, and the intermediate coordinate frame, $H_i-x'_iy'_iz'_i$, attached at point H_i. Let $\mathbf{X}^i$, $\mathbf{X}'$ and $\mathbf{X}^{i\text{-}1}$ be 4×1 position vectors in $O_i-x_iy_iz_i$, $H_i-x'y'z'$, and $O_{i-1}-x_{i-1}y_{i-1}z_{i-1}$, respectively. From the figure, the coordinate transformation from $\mathbf{X}^i$ to $\mathbf{X}'$ is given by

$$\mathbf{X}' = \mathbf{A}_i^{int}\mathbf{X}^i \qquad (2-31)$$

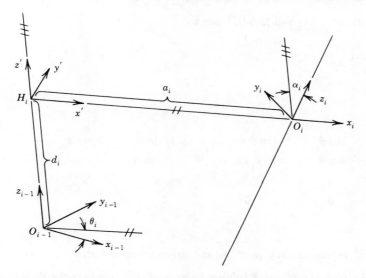

Figure 2-10 :   The relationship between adjacent coordinate frames in the Denavit-Hartenberg notation.

where

$$
\mathbf{A}_i^{int} =
\begin{bmatrix}
1 & 0 & 0 & a_i \\
0 & \cos \alpha_i & -\sin \alpha_i & 0 \\
0 & \sin \alpha_i & \cos \alpha_i & 0 \\
0 & 0 & 0 & 1
\end{bmatrix}
\tag{2-32}
$$

Similarly the transformation from $\mathbf{X'}$ to $\mathbf{X}^{i-1}$ is given by

$$
\mathbf{X}^{i-1} = A_{int}^{i-1} \, \mathbf{X'}
\tag{2-33}
$$

where

$$
\mathbf{A}_{int}^{i-1} =
\begin{bmatrix}
\cos \theta_i & -\sin \theta_i & 0 & 0 \\
\sin \theta_i & \cos \theta_i & 0 & 0 \\
0 & 0 & 1 & d_i \\
0 & 0 & 0 & 1
\end{bmatrix}
\tag{2-34}
$$

Combining equations (2-31) and (2-33), leads to

$$\mathbf{X}^{i-1} = \mathbf{A}_i^{i-1} \, \mathbf{X}^i \qquad\qquad (2\text{--}33)$$

where

$$\mathbf{A}_i^{i-1} = \begin{bmatrix} \cos\theta_i & -\sin\theta_i\cos\alpha_i & \sin\theta_i\sin\alpha_i & a_i\cos\theta_i \\ \sin\theta_i & \cos\theta_i\cos\alpha_i & -\cos\theta_i\sin\alpha_i & a_i\sin\theta_i \\ 0 & \sin\alpha_i & \cos\alpha_i & d_i \\ 0 & 0 & 0 & 1 \end{bmatrix} \qquad (2\text{--}35)$$

The matrix $\mathbf{A}_i^{i-1}$ represents the position and orientation of frame i relative to frame i-1. As shown before, the first three 3×1 column vectors of A_i^{i-1} contain the direction cosines of the coordinate axes of frame i, while the last 3×1 column vector represents the position of the origin O_i.

2.2.3. Kinematic Equations

Using the Denavit-Hartenberg notation we express the position and orientation of the end-effector as a function of joint displacements in this section. The displacement of each joint is either angle θ_i or distance d_i, depending on the type of joint. In general we denote the joint displacement by q_i, which is defined as

$$q_i = \theta_i \qquad\qquad \text{for a revolute joint}$$

$$q_i = d_i \qquad\qquad \text{for a prismatic joint}$$

The position and orientation of link i relative to link i-1 is then described as a function of q_i using the 4×4 matrix $A_i^{i-1}(q_i)$.

Our primary goal in this section is to describe the position and orientation of the last link with reference to the base frame, as a function of joint displacements, q_1 through q_n. As shown in Figure 2-11, the manipulator arm consists of $(n+1)$ links from the base to the tip, in

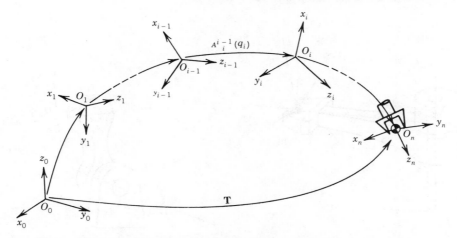

Figure 2-11 :   The representation of the end-effector location by a 4x4 matrix.

which relative locations of adjacent links are represented by the 4×4 matrices. Considering the n consecutive coordinate transformations along the serial linkage, we can derive the end-effector location viewed from the base frame.   Namely, from (2-27), the position and orientation of the last link relative to the base frame is given by

$$\mathbf{T} = \mathbf{A}_1^0(q_1)\mathbf{A}_2^1(q_2) \; \cdots \; \mathbf{A}_n^{n-1}(q_n) \tag{2-36}$$

where $\mathbf{T}$ is a 4×4 matrix representing the position and orientation of the last link with reference to the base frame, as shown in Figure 2-11.   Equation (2-36) provides the functional relationship between the last link position and orientation and the displacements of all the joints involved in the open kinematic chain.   It is referred to as the *kinematic equation* of the manipulator arm, and governs the fundamental kinematic behavior of the arm.

There are several exceptions to the Denavit-Hartenberg notation rule.   To define a coordinate frame attached to each link, the common normal between the two joint axes must be determined for the link.   However, no such common normals exist for the base and the last links, since each of these links has only one joint axis.   For these two links, the coordinate frames are defined as follows.   For the last link, the origin of the coordinate frame can be

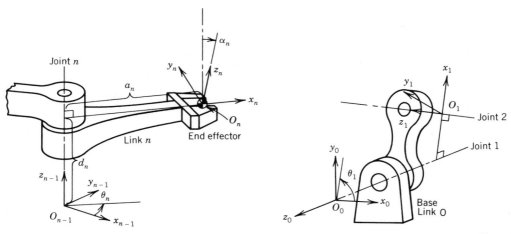

Figure 2-12 : Location of the end coordinate frame **Figure 2-13** : Location of the base coordinate frame.

chosen at any convenient point of the end-effector, as shown in Figure 2-12. The orientation of the coordinate frame, however, must be determined so that the x_n axis intersects the last joint axis at a right angle. The angle α_n in the figure is arbitrary. For the base link, the origin of the coordinate frame can be chosen at an arbitrary point on the joint axis 1, as shown in Figure 2-13. The z_0 axis must be parallel to the joint axis, while the orientation of the x and y axes about the joint axis is arbitrary.

Further, there are two exceptions to note for the intermediate links between the base and the last links. When the two joint axes of an intermediate link are parallel, the common normal between them is not unique. The choice of common normal is then arbitrary. Usually, one chooses the common normal that passes through O_{i-1} in Figure 2-9, so that the distance d_i becomes zero. The other exception concerns prismatic joints. For a prismatic joint, only the direction of the joint axis is meaningful, hence the position of the joint axis can be chosen arbitrarily.

Let us work out an example to become familiar with kinematic modeling.

Example 2-4. The Kinematic Model of a 5-R-1-P Manipulator Arm.

Figure 2-14 shows a six-degree-of-freedom manipulator arm with five revolute joints and one prismatic joint, or *5-R-1-P* manipulator arm. This type of arm structure has been used

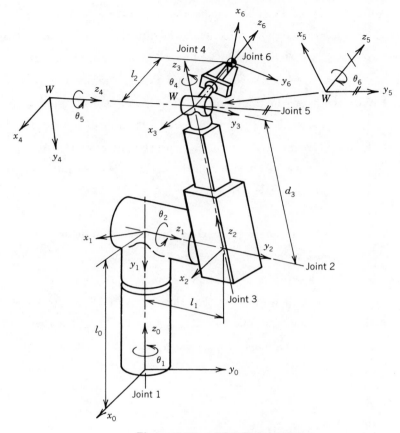

Figure 2-14 :   5-R-1-P manipulator.

widely for commercial robots and research manipulators. Let us derive the kinematic equation for this arm.

The first step of kinematic modeling is to identify all the joints. The first joint, joint 1, is a revolute joint, which rotates the whole body about the vertical axis. Joint 2 is also a revolute joint about the horizontal axis. Joint 3 is a prismatic joint that causes translational motions of the last three links. Here, the position of the joint axis of joint 3 has been chosen so that it coincides with joint 4. The last three joints are all revolute joints, and their axes intersect at the single point W shown in the figure.

Next, coordinate frames must be attached to the arm links. The base frame is chosen to be on the table surface with the z_0 axis along the joint axis. Since joint axes 1 and 2 intersect as shown in the figure, the length of the common normal is zero, and it is directed along the perpendicular to both joint axes. This direction is the direction of the x_1 axis, according to the Denavit-Hartenberg notation. The second coordinate frame is also at the intersection of joint axes 2 and 3. Since joint axes 3 and 4 coincide, their common normal is not unique and can be chosen arbitrarily on the axes. In the figure, the third coordinate frame has been chosen at point W so that the three axes of the frame are parallel to those of the second frame. Frames 4 and 5 are also located at point W, because the joint axes intersect at this point. The origin of the final frame can be selected arbitrarily. In the figure, it is chosen at an appropriate point on the last joint axis at which a workpiece will be grasped.

The Denavit-Hartenberg parameters for these coordinate frames are listed in Table 2-1. Note that many parameters in the table are equal to zero. We have defined the coordinate frames so that the kinematic equation be simple and only include a small number of non-zero parameters. The table also shows that the joint variable for joint 3 is displacement d_3 and that all the other joint variables are angles θ_i. The 4×4 matrix $A_i^{i-1}(q_i)$ can be determined by substituting the parameters listed in the table into equation (2-36):

$$\mathbf{A}_1^0(\theta_1) = \begin{bmatrix} c_1 & 0 & -s_1 & 0 \\ s_1 & 0 & c_1 & 0 \\ 0 & -1 & 0 & l_0 \\ 0 & 0 & 0 & 1 \end{bmatrix} \qquad \mathbf{A}_2^1(\theta_2) = \begin{bmatrix} c_2 & 0 & s_2 & 0 \\ s_2 & 0 & -c_2 & 0 \\ 0 & 1 & 0 & l_1 \\ 0 & 0 & 0 & 1 \end{bmatrix}$$

$$\mathbf{A}_3^2(d_3) = \begin{bmatrix} 1 & 0 & 0 & 0 \\ 0 & 1 & 0 & 0 \\ 0 & 0 & 1 & d_3 \\ 0 & 0 & 0 & 1 \end{bmatrix} \qquad \mathbf{A}_4^3(\theta_4) = \begin{bmatrix} c_4 & 0 & -s_4 & 0 \\ s_4 & 0 & c_4 & 0 \\ 0 & -1 & 0 & 0 \\ 0 & 0 & 0 & 1 \end{bmatrix} \qquad (2\text{-}38)$$

$$\mathbf{A}_5^4(\theta_5) = \begin{bmatrix} c_5 & 0 & s_5 & 0 \\ s_5 & 0 & -c_5 & 0 \\ 0 & 1 & 0 & 0 \\ 0 & 0 & 0 & 1 \end{bmatrix} \qquad \mathbf{A}_6^5(\theta_6) = \begin{bmatrix} c_6 & -s_6 & 0 & 0 \\ s_6 & c_6 & 0 & 0 \\ 0 & 0 & 1 & l_2 \\ 0 & 0 & 0 & 1 \end{bmatrix}$$

where $c_i = \cos(\theta_i)$ and $s_i = \sin(\theta_i)$. The kinematic equation of this manipulator arm is then

Table 2-1 : Link parameters for the 5-R-1-P manipulator.

link number	α_i	a_i	d_i	θ_i
1	-90°	0	l_0	θ_1
2	+90°	0	l_1	θ_2
3	0	0	d_3	0
4	-90°	0	0	θ_4
5	+90°	0	0	θ_5
6	0	0	l_2	θ_6

given by

$$\mathbf{T} = \mathbf{A}_1^0(\theta_1)\mathbf{A}_2^1(\theta_2)\mathbf{A}_3^2(d_3)\mathbf{A}_4^3(\theta_4)\mathbf{A}_5^4(\theta_5)\mathbf{A}_6^5(\theta_6) \qquad (2-39)$$

The end-effector position and orientation $\mathbf{T}$ is represented as a function of joint displacements, θ_1 ,θ_2 ,d_3 ,θ_4 , θ_5 , and θ_6 . ▵▵▵

2.3. Inverse Kinematics

2.3.1. Introduction

The kinematic equation derived in the previous section provides the functional relationship between the joint displacements and the resultant end-effector position and orientation. By substituting values of joint displacements into the right-hand side of the kinematic equation, one can immediately find the corresponding end-effector position and orientation. The problem of finding the end-effector position and orientation for a given set of joint displacements is referred to as the *direct kinematics problem*. This is simply to evaluate the right-hand side of the kinematic equation for known joint displacements.

In this section, we discuss the problem of moving the end-effector of a manipulator arm to a specified position and orientation. We need to find the joint displacements that lead the end-effector to the specified position and orientation. This is the inverse of the previous problem, and is thus referred to as the *inverse kinematics problem*. The kinematic equation must be solved for joint displacements, given the end-effector position and orientation. Once the kinematic equation is solved, the desired end-effector motion can be achieved by moving each joint to the determined value.

In the direct kinematics problem, the end-effector location is determined uniquely for any given set of joint displacements. On the other hand, the inverse kinematics is more complex in the sense that multiple solutions may exist for the same end-effector location. Also, solutions may not always exist for a particular range of end-effector locations and arm structures. Further, since the kinematic equation is comprised of nonlinear simultaneous equations with many trigonometric functions, it is not always possible to derive a closed-form solution, which is the explicit inverse function of the kinematic equation. When the kinematic equation cannot be solved analytically, numerical methods are used in order to derive the desired joint displacements.

A manipulator arm must have at least six degrees of freedom in order to locate its end-effector at an arbitrary point with an arbitrary orientation in space. Manipulator arms with less than 6 degrees of freedom are not able to perform such arbitrary positioning. On the other hand, if a manipulator arm has more than 6 degrees of freedom, there exist an infinite number of solutions to the kinematic equation. Consider for example the human arm, which has seven degrees of freedom, if we exclude the joints at the fingers. Even if the hand is fixed on a table, one can change the elbow position continuously without changing the hand location. This implies that there exist an infinite set of joint displacements that lead the hand to the same location. Manipulator arms with more than six degrees of freedom are referred to as *redundant manipulators*. We will discuss redundant manipulators in detail in the following chapter. In this chapter we focus on inverse kinematics for six degree-of-freedom manipulator arms.

2.3.2. Solving the Kinematic Equation for the 5-R-1-P Manipulator

In this section we solve the kinematic equation that we obtained for the 5-R-1-P manipulator of Example 2-4. The kinematic equation was given by

$$\mathbf{T} = \mathbf{A}_1^0 \, \mathbf{A}_2^1 \, \mathbf{A}_3^2 \, \mathbf{A}_4^3 \, \mathbf{A}_5^4 \, \mathbf{A}_6^5 \tag{2-39}$$

For this manipulator arm, closed-form solutions exist for an arbitrary end-effector location $\mathbf{T}$. The above equation can be written in many different forms. For example, postmultiplying both sides by the inverse of $\mathbf{A}_6^5$ yields

$$\mathbf{T}(\mathbf{A}_6^5)^{-1} = \mathbf{A}_1^0 \, \mathbf{A}_2^1 \, \mathbf{A}_3^2 \, \mathbf{A}_4^3 \, \mathbf{A}_5^4 \tag{2-40}$$

Further premultiplying both sides by $(\mathbf{A}_1^0)^{-1}$,

$$(\mathbf{A}_1^0)^{-1}\mathbf{T}(\mathbf{A}_6^5)^{-1} = \mathbf{A}_2^1 \, \mathbf{A}_3^2 \, \mathbf{A}_4^3 \, \mathbf{A}_5^4 \tag{2-41}$$

The left-hand side of equation (2-40) is only a function of θ_6 , while the right-hand side involves all the other joint displacements. Similarly, equation (2-41) has θ_1 and θ_6 on the left-hand side and the remaining joint displacements on the right-hand side. We need to find an appropriate expression that permits us to solve the kinematic equation conveniently.

To this end, let us interpret the physical meanings of the different expressions by using a graphic representation. Figure 2-15 shows the skeleton structure of the 5-R-1-P manipulator. Each arc in the figure represents the relationship between the two coordinate frames, and the 4×4 matrix on the arc gives the position and orientation of the frame viewed from the frame at the origin of the arc. The product of multiple matrices represents the position and orientation of the final frame viewed from the initial frame along the path of the arcs associated with the matrices. The left-hand side of the original kinematic equation represents the end-effector position and orientation viewed from the base frame directly, while the right-hand side represents the same end-effector position and orientation through another path along the arm linkage. Both sides of equation (2-40) represent the position and orientation of the frame attached to link 5 with reference to the base frame through two different paths reaching the same frame. The origin of the coordinate frame 5 is at point W, the coordinates of which are represented by the fourth column of the 4×4 matrices in equation (2-40). Note, however, that the position of W in the figure depends only on the first three joints, and is independent of the last three joints. Therefore, if one compares the fourth column vectors of the matrices

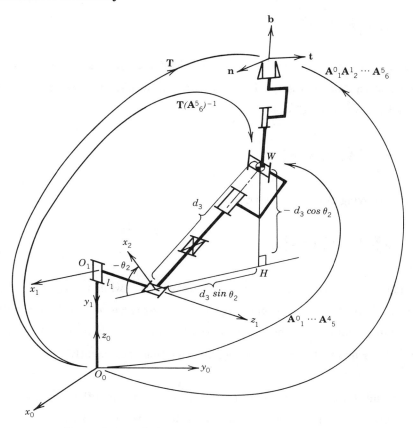

Figure 2-15 :   Skeleton structure of the 5-R-1-P manipulator.

on both sides of (2-40), simultaneous equations with only three unknowns should be obtained. Further, a more convenient form of simultaneous equations can be derived by evaluating the fourth columns of equation (2-41). The fourth column vector of the right-hand side of (2-41) represents the position of W with respect to the first coordinate frame through the arm linkage, as shown in Figure 2-15, and is simply given by

$$\mathbf{x}_W^1 = \begin{pmatrix} d_3 s_2 \\ -d_3 c_2 \\ l_1 \end{pmatrix} \tag{2-42}$$

The left-hand side of (2-41) describes the same position, now reached through the base frame and the end-effector. Thus, writing the desired end-effector position and orientation $\mathbf{T}$ in the form

$$\mathbf{T} = \begin{bmatrix} n_x & t_x & b_x & p_x \\ n_y & t_y & b_y & p_y \\ n_z & t_z & b_z & p_z \\ 0 & 0 & 0 & 1 \end{bmatrix} \tag{2-43}$$

and substituting into the left-hand side of equation (2-41), we obtain another expression of coordinates $\mathbf{x}_W^1$, namely

$$\mathbf{x}_W^1 = \begin{bmatrix} \overset{*}{p_x}c_1 + \overset{*}{p_y}s_1 \\ -\overset{*}{p_z} + l_0 \\ -\overset{*}{p_x}s_1 + \overset{*}{p_y}c_1 \end{bmatrix} \tag{2-44}$$

where $\overset{*}{p_x}, \overset{*}{p_y}, \overset{*}{p_z}$ represent the coordinates of point W, and are given by

$$\begin{aligned} \overset{*}{p_x} &= p_x - l_2 b_x \\ \overset{*}{p_y} &= p_y - l_2 b_y \\ \overset{*}{p_z} &= p_z - l_2 b_z \end{aligned} \tag{2-45}$$

Equating (2-42) and (2-44) yields three equations with three unknowns:

$$d_3 s_2 = \overset{*}{p_x}c_1 + \overset{*}{p_y}s_1 \tag{2-46}$$

$$-d_3 c_2 = -\overset{*}{p_z} + l_0 \tag{2-47}$$

$$l_1 = -\overset{*}{p_x}s_1 + \overset{*}{p_y}c_1 \tag{2-48}$$

To solve the last equation, we let:

$$t = \tan\left(\frac{\theta_1}{2}\right) \tag{2-49}$$

so that

$$c_1 = \cos\theta_1 = \frac{1-t^2}{1+t^2} \qquad \text{and} \qquad s_1 = \sin\theta_1 = \frac{2t}{1+t^2} \tag{2-50}$$

Substituting expressions (2-50) into equation (2-48), we obtain a quadratic equation in terms of the unknown variable t :

$$(l_1 + p_y^*)t^2 + 2p_x^*t + l_1 - p_y^* = 0 \tag{2-51}$$

Solving the above equation for t and using (2-49) yields

$$\theta_1 = 2\tan^{-1}\left[\frac{-p_x^* \pm \sqrt{p_x^{*2} + p_y^{*2} - l_1^2}}{l_1 + p_y^*}\right] \tag{2-52}$$

Note that the quantity under the square root must be positive. Otherwise, the solution does not exist, which means that the specified end-effector position is out of the reachable range, or *workspace*, of the manipulator arm.

Dividing both sides of (2-46) by the corresponding sides of (2-47), we obtain

$$\theta_2 = \tan^{-1}\left[\frac{p_x^*c_1 + p_y^*s_1}{p_z^* - l_0}\right] \tag{2-53}$$

Using (2-52) in the above expression allows us to evaluate the unknown angle θ_2 . Further, d_3 can be obtained by taking the sum of the squares of equations (2-46) and (2-47) :

$$d_3 = \pm\sqrt{(p_x^*c_1 + p_y^*s_1)^2 + (p_z^* - l_0)^2} \tag{2-54}$$

Let us discuss the solutions. The prismatic joint shown in Figure 2-14 has a limited stroke in which d_3 is always positive. Therefore, the negative solution in equation (2-54) must be eliminated. Equation (2-52), on the other hand, has two solutions due to the double signs before the square root. Also, expression (2-53) takes two different values depending on the

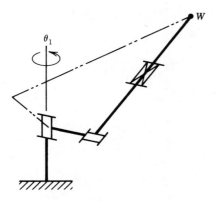

Figure 2-16 :   Multiple configurations of the shoulder joints.

value of θ_1. Figure 2-16 illustrates the two configurations of the manipulator arm corresponding to the two solutions. These two configurations yield the same point W. Thus inverse kinematics solutions are generally not unique. Note also that the arctangent functions in equations (2-52) and (2-53) can take two values, which are 180 degrees apart. An appropriate quadrant must be chosen for each equation depending on the signs of both the numerator and the denominator of the arctangent function.

After the first three joint displacements are determined, we solve the kinematic equation for the last three joint displacements. Premultiplying the inverse of the matrix product $\mathbf{A}_1^0\mathbf{A}_2^1\mathbf{A}_3^2$ to both sides of equation (2-38), we obtain

$$\left[\mathbf{A}_1^0(\theta_1)\mathbf{A}_2^1(\theta_2)\mathbf{A}_3^2(d_3)\right]^{-1}\mathbf{T} = \mathbf{A}_4^3(\theta_4)\mathbf{A}_5^4(\theta_5)\mathbf{A}_6^5(\theta_6) \qquad (2\text{--}55)$$

Both sides of this equation represent the position and orientation of the end-effector viewed from the third frame. Since θ_1 , θ_2 , and d_3 have been determined, the left-hand side matrix is known. Let us denote it by

$$\mathbf{T}' = [\mathbf{A}_1^0\mathbf{A}_2^1\mathbf{A}_3^2]^{-1}\mathbf{T} = \begin{bmatrix} n_{x'} & t_{x'} & b_{x'} & p_{x'} \\ n_{y'} & t_{y'} & b_{y'} & p_{y'} \\ n_{z'} & t_{z'} & b_{z'} & p_{z'} \\ 0 & 0 & 0 & 1 \end{bmatrix} \qquad (2\text{--}56)$$

Premultiplying equation (2-55) by $\left[\mathbf{A}_4^3(\theta_4)\right]^{-1}$ and evaluating both sides, we get

$$
(\mathbf{A}_4^3)^{-1}\mathbf{T}' = \begin{bmatrix} * & * & b_{x'}c_4+b_{y'}s_4 & * \\ -n_{z'} & -t_{z'} & -b_{z'} & * \\ -n_{x'}s_4+n_{y'}c_4 & -t_{x'}s_4+t_{y'}c_4 & -b_{x'}s_4+b_{y'}c_4 & * \\ 0 & 0 & 0 & 1 \end{bmatrix} \qquad (2\text{--}57)
$$

$$
\mathbf{A}_5^4\mathbf{A}_6^5 = \begin{bmatrix} c_5c_6 & -c_5s_6 & s_5 & * \\ s_5c_6 & -s_5s_6 & -c_5 & * \\ s_6 & c_6 & 0 & * \\ 0 & 0 & 0 & 1 \end{bmatrix} \qquad (2\text{--}58)
$$

where some elements of the matrices are simply denoted by * , since they are irrelevant in the present calculation. Comparing the [3,3] elements of the above matrices, we obtain

$$
-b_{x'}s_4 + b_{y'}c_4 = 0 \qquad (2\text{--}59)
$$

Joint displacement θ_4 is then given by

$$
\theta_4 = \tan^{-1}\!\left(\frac{b_{y'}}{b_{x'}}\right) \qquad (2\text{--}60)
$$

From the [1,3] and [2,3] elements, we get

$$
\theta_5 = \tan^{-1}\!\left(\frac{b_{x'}c_4 + b_{y'}s_4}{b_{z'}}\right) \qquad (2\text{--}61)
$$

where c_4 and s_4 are evaluated with equation (2-60). Similarly joint displacement θ_6 can be determined from the elements [3,1] and [3,2], and given by

$$
\theta_6 = \tan^{-1}\!\left(\frac{-n_{x'}s_4 + n_{y'}c_4}{-t_{x'}s_4 + t_{y'}c_4}\right) \qquad (2\text{--}62)
$$

Thus all six joint displacements are obtained.

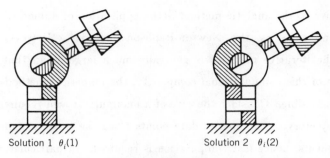

Solution 1 $\theta_i(1)$ Solution 2 $\theta_i(2)$

Figure 2-17 :   Multiple configurations of the wrist joints.

If each joint is allowed to rotate 360 degrees, there are two possible solutions for the last three joints displacements. Indeed, since equation (2-60) involves the arctangent function, the angle θ_4 may have two values, which are 180 degrees apart. The two configurations corresponding to the two solutions are illustrated in Figure 2-17. Let $\theta_i(1)$ and $\theta_i(2)$ be the first and the second solutions, with $\theta_i(1) \leq \theta_i(2)$. These solutions are related by

$$\theta_4(2) = \theta_4(1) + \pi$$

$$\theta_5(2) = -\theta_5(1) \qquad\qquad\qquad (2-63)$$

$$\theta_6(2) = \theta_6(1) + \pi$$

The Euler angles discussed in Section 2.1.3 were defined for the three consecutive rotations where the three axes of rotations intersect at a single point. The last three joints of the 5-R-1-P manipulator have the same construction as that of the Euler angles. Note that the Euler angles are not uniquely determined for a given orientation, as shown by equation (2-63).

2.3.3. Solvability

In the previous section, we solved the kinematic equation for the 5-R-1-P manipulator completely, and obtained closed-form solutions that provide the joint displacements as explicit functions in terms of the desired position and orientation of the end-effector. The joint displacements can be determined by simply evaluating these functions for the desired end-effector location data. On the other hand, if the closed-form solutions cannot be obtained, we cannot find joint displacements in such a straightforward, analytic way.

An alternative to the analytic method is the application of numerical methods based on iterative algorithms such as the Newton-Raphson method. However, the amount of computation for the iterative methods is generally much larger than that of the close-form solutions. Because of this computational complexity, the numerical methods often turn out to be impractical or infeasible. Consider the case of a manipulator arm required to move its end-effector along a trajectory. A number of data points along the trajectory must be transformed into joint displacements, hence fast computation is required. In particular, computation time is crucial if the transformation must be performed in real time.

The existence of a closed-form solution depends on the kinematic structure of the manipulator arm. For an appropriate structure such as that of the 5-R-1-P manipulator arm, for instance, a closed-form solution can be obtained. A kinematic structure for which a closed-form solution exists is referred to as a *solvable structure*. The kinematic structure of a manipulator arm is often designed so that the kinematic equation be solvable, in order to avoid computational complexity. Most industrial robots do have solvable structures.

Thus, an important issue is to find what makes a kinematic structure solvable. (Pieper, 1968) shows that a sufficient condition for the kinematic structure of a six degree-of-freedom manipulator arm to be solvable is that the joint axes of three consecutive revolute joints intersect at a single point for all the arm configurations. The 5-R-1-P manipulator discussed previously does satisfy the sufficient condition, since the axes of the last three joints intersect at the single point W, as shown in Figure 2-15.

When the last three joint axes intersect at a single point, the three joints are often referred to as a *spherical wrist*. (Pieper, 1968) lists all the possible kinematic structures for six degree-of-freedom manipulators in which the axes of three consecutive revolute joints intersect at a single point, and closed-form solutions to each of the structures are obtained analytically. To simply illustrate the proof, let us recall the process of solving the 5-R-1-P kinematic equation. First, we looked at the point W at which the three revolute joint axes intersect. By rewriting the original kinematic equation so that both sides represent the coordinates of the point W, we divided the original problem of finding six unknowns into two problems with only three unknowns each. This division is always possible if the axes of three consecutive joints intersect at a single point. For the divided simultaneous equations with three unknowns, it can be shown that there exist closed-form solutions. Comparing the corresponding elements of the matrix equation and using an appropriate change of variables such as (2-49), we can reduce the equation into an algebraic equation of order at most four, for which analytical solutions exist.[1] Therefore, closed-form solutions can be obtained. This procedure for solving the kinematic equation is not applicable in general when the axes of three consecutive revolute joints do not meet at a single point. Note, however, that the solvability condition given by Pieper is not necessary but only sufficient. Therefore, other types of kinematic structures may also be solvable.

2.4. Research Topics

Numerous kinematic modeling methods can be used as alternatives to the homogenous transformations. (Yang and Freudenstein, 1964), for instance, apply *dual-number quaternion* algebra to the kinematic modeling of spatial mechanisms. The method allows one to treat rigid body translations and rotations in a simple and convenient manner, and is increasingly used in manipulator kinematics (Roth, 1983; Pennock and Yang, 1985-a; Sugimoto, 1984; Waldron *et al.*, 1985).

On the basis of the kinematic modeling and analysis, the design of arm linkage has been dealt with extensively. (Roth, 1975) was the first to address the design problem of finding appropriate kinematic structures and link dimensions to allow the arm to cover a specified workspace. Since then, a variety of analytic and numerical methods were presented to study the shape and volume of the workspace (Shimano, 1978; Kumar and Waldron, 1981; Sugimoto and Duffy, 1981; Gupta and Roth, 1982; Tsai and Soni, 1981; Kohli and Spanos, 1985; Cwiakala and Lee, 1985; Yang and Lai, 1985).

The inverse kinematics problem of general manipulator arms has also been discussed extensively. Numerical methods based upon iterative computation algorithms were devised by e.g. (Whitney, 1969-b; Tsai and Morgan, 1985). Pieper's analytic method was extended by (Featherstone, 1983-b; Hollerbach, 1983).

---

[1]Algebraic equations of order higher than four are not solvable analytically.

Chapter 3
KINEMATICS II: DIFFERENTIAL MOTION

In the previous chapter, the position and orientation of the manipulator end-effector were evaluated in relation to joint displacements. The joint displacements corresponding to a given end-effector location were obtained by solving the kinematic equation for the manipulator. This preliminary analysis permitted the manipulator to place the end-effector at a specified location in space.

In this chapter, we are concerned not only with the final location of the end-effector, but also with the *velocity* at which the end-effector moves. In order to move the end-effector in a specified direction at a specified speed, it is necessary to *coordinate* the motion of the individual joints. The focus of this chapter is the development of fundamental methods for achieving such coordinated motion in multiple joint manipulators. As discussed in the previous chapter, the end-effector position and orientation are directly related to the joint displacements; hence, in order to coordinate joint motions, we derive the *differential* relationship between the joint displacements and the end-effector location, and then solve for the individual joint motions.

3.1. Kinematic Modeling of Instantaneous Motions

3.1.1. Differential Relationships

We begin by considering the two degree-of-freedom planar manipulator shown in Figure 3-1. Here, the manipulator arm is constrained to the $x_0 - y_0$ plane. The kinematic equations relating the end-effector position (x, y) to the joint displacements (θ_1, θ_2) are given by

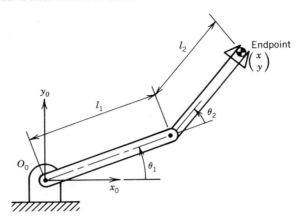

Figure 3-1 :   Two degree-of-freedom planar manipulator.

$$x(\theta_1, \theta_2) = l_1 \cos \theta_1 + l_2 \cos (\theta_1 + \theta_2)$$

$$y(\theta_1, \theta_2) = l_1 \sin \theta_1 + l_2 \sin (\theta_1 + \theta_2)$$

$$(3-1)$$

In this chapter, we are interested in the small motions of the end-effector about its current position. The infinitesimal motion relationship is determined by differentiating the kinematic equations. For two degree-of-freedom planar manipulation, the differential relationship is obtained by simply differentiating equation (3-1) so that

$$dx = \frac{\partial x(\theta_1, \theta_2)}{\partial \theta_1} d\theta_1 + \frac{\partial x(\theta_1, \theta_2)}{\partial \theta_2} d\theta_2$$

$$(3-2)$$

$$dy = \frac{\partial y(\theta_1, \theta_2)}{\partial \theta_1} d\theta_1 + \frac{\partial y(\theta_1, \theta_2)}{\partial \theta_2} d\theta_2$$

In vector form the above can be written as

$$d\mathbf{x} = \mathbf{J} d\boldsymbol{\theta}$$

$$(3-3)$$

where $d\mathbf{x}$ and $d\boldsymbol{\theta}$ are infinitesimal displacement vectors defined as

$$d\mathbf{x} = \begin{bmatrix} dx \\ dy \end{bmatrix} \qquad\qquad d\boldsymbol{\theta} = \begin{bmatrix} d\theta_1 \\ d\theta_2 \end{bmatrix}$$

$$(3-4)$$

and

$$
\mathbf{J} = \begin{bmatrix} \dfrac{\partial x}{\partial \theta_1} & \dfrac{\partial x}{\partial \theta_2} \\[2ex] \dfrac{\partial y}{\partial \theta_1} & \dfrac{\partial y}{\partial \theta_2} \end{bmatrix}
\qquad (3\text{--}5)
$$

The matrix $\mathbf{J}$ comprises the partial derivatives of the functions $x(\theta_1,\theta_2)$ and $y(\theta_1,\theta_2)$ with respect to joint displacements θ_1 and θ_2. The matrix $\mathbf{J}$ is referred to as the *manipulator Jacobian*. The manipulator Jacobian represents the infinitesimal relationship between the joint displacements and the end-effector location at the present position and arm configuration.

From equation (3-1), the Jacobian matrix of the two degree-of-freedom planar manipulator is given by

$$
\mathbf{J} = \begin{bmatrix} -l_1\sin\theta_1 - l_2\sin(\theta_1+\theta_2) & -l_2\sin(\theta_1+\theta_2) \\[1ex] l_1\cos\theta_1 + l_2\cos(\theta_1+\theta_2) & l_2\cos(\theta_1+\theta_2) \end{bmatrix}
\qquad (3\text{--}6)
$$

Note that the elements of the Jacobian are functions of joint displacements, and therefore vary with the arm configuration.

Consider the instant when the two joints of the two degree-of-freedom planar manipulator move at joint velocities $\dot{\boldsymbol{\theta}} = [\dot\theta_1,\dot\theta_2]^T$, and let $\mathbf{v} = [\dot x, \dot y]^T$ be the resulting end-effector velocity vector. The Jacobian represents the relationship between the joint velocities and the resulting end-effector velocities as well as the infinitesimal position relationship. Indeed, dividing both sides of (3-3) by the infinitesimal time increment dt, we obtain

$$
\frac{d\mathbf{x}}{dt} = \mathbf{J}\frac{d\boldsymbol{\theta}}{dt}
$$

that is,

$$
\mathbf{v} = \mathbf{J}\dot{\boldsymbol{\theta}}
\qquad (3\text{--}7)
$$

Thus the velocity relationship between the joints and the end-effector is determined by the manipulator Jacobian.

Let $\mathbf{J}_1$ and $\mathbf{J}_2$ be two 2×1 vectors consisting respectively of the first and the second columns of the Jacobian (3-6). Equation (3-7) can then be rewritten as

$$\mathbf{v} = \mathbf{J}_1 \dot{\theta}_1 + \mathbf{J}_2 \dot{\theta}_2 \qquad (3-8)$$

The first term on the right-hand side accounts for the end-effector velocity induced by the first joint only, while the second term represents the velocity resulting from the second joint motion only. The resultant end-effector velocity is given by the vector sum of the two. Each column vector of the Jacobian matrix represents the end-effector velocity generated by the corresponding joint motion at unit velocity when all other joints are immobilized.

3.1.2. Infinitesimal Rotations

In the previous section we dealt with simple planar manipulation and analyzed the infinitesimal translation and the linear velocity of the endpoint. To generalize the result, we need to include the infinitesimal rotation and angular velocity for a general spatial manipulator arm.

In Chapter 2, we developed mathematical tools for representing the spatial orientation of a rigid body. Those methods utilize 3×3 rotation matrices and Euler angles and allow us to represent rotations and orientations of finite angles. However, infinitesimal rotations or time derivatives of orientations are substantially different from finite angle rotations and orientations. As a result, the methods for representing finite rotations and orientations are not appropriate when infinitesimal motions are considered. In this section, we investigate the difference between finite and infinitesimal rotations, and then develop an appropriate mathematical tool for infinitesimal rotations.

We begin by writing the 3×3 rotation matrix representing infinitesimal rotation $d\phi_x$ about the x axis:

$$\mathbf{R}_x(d\phi_x) = \begin{bmatrix} 1 & 0 & 0 \\ 0 & \cos(d\phi_x) & -\sin(d\phi_x) \\ 0 & \sin(d\phi_x) & \cos(d\phi_x) \end{bmatrix} \approx \begin{bmatrix} 1 & 0 & 0 \\ 0 & 1 & -d\phi_x \\ 0 & d\phi_x & 1 \end{bmatrix} \qquad (3-9)$$

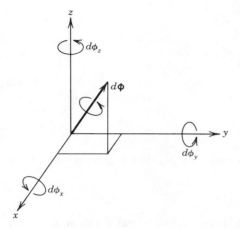

Figure 3-2 : Infinitesimal rotation vector.

Note that, since $d\phi_x$ is infinitesimal, $\cos(d\phi_x)=1$ and $\sin(d\phi_x)=d\phi_x$. For infinitesimal rotations about the y and z axes, similar matrix expressions can be obtained in the same way as equation (3-9). Let $\mathbf{R}_y(d\phi_y)$ be the 3×3 infinitesimal rotation matrix about the y axis; then the result of consecutive rotations about the x and y axes is given by

$$
\mathbf{R}_x(d\phi_x)\mathbf{R}_y(d\phi_y) =
\begin{bmatrix}
1 & 0 & 0 \\
0 & 1 & -d\phi_x \\
0 & d\phi_x & 1
\end{bmatrix}
\begin{bmatrix}
1 & 0 & d\phi_y \\
0 & 1 & 0 \\
-d\phi_y & 0 & 1
\end{bmatrix}
$$

$$
=
\begin{bmatrix}
1 & 0 & d\phi_y \\
d\phi_x d\phi_y & 1 & -d\phi_x \\
-d\phi_y & d\phi_x & 1
\end{bmatrix}
=
\begin{bmatrix}
1 & 0 & d\phi_y \\
0 & 1 & -d\phi_x \\
-d\phi_y & d\phi_x & 1
\end{bmatrix}
$$

$$(3-10)$$

where the higher order quantity $d\phi_x d\phi_y$ is neglected. We now change the order of rotations $\mathbf{R}_y(d\phi_y)$ and $\mathbf{R}_x(d\phi_x)$. Calculating the matrix product in the same way as before, we find that the two results are identical. Namely,

$$
\mathbf{R}_x(d\phi_x)\mathbf{R}_y(d\phi_y) = \mathbf{R}_y(d\phi_y)\mathbf{R}_x(d\phi_x)
$$

$$(3-11)$$

Therefore, infinitesimal rotations do not depend on the order of rotations; in other words, they *commute*. In general, infinitesimal rotations about the x, y, and z axes shown in Figure 3-2 can be represented by

$$\mathbf{R}(d\phi_x, d\phi_y, d\phi_z) = \begin{bmatrix} 1 & -d\phi_z & d\phi_y \\ d\phi_z & 1 & -d\phi_x \\ -d\phi_y & d\phi_x & 1 \end{bmatrix} \tag{3-12}$$

The rotation matrix depends only on the three inifinitesimal angles, but is independent of the order of rotations.

Let $\mathbf{R}(d\phi_x, d\phi_y, d\phi_z)$ and $\mathbf{R}(d\phi'_x, d\phi'_y, d\phi'_z)$ be two infinitesimal rotation matrices, then the consecutive rotations of the two yield

$$\mathbf{R}(d\phi_x, d\phi_y, d\phi_z)\mathbf{R}(d\phi'_x, d\phi'_y, d\phi'_z)$$

$$= \begin{bmatrix} 1 & (d\phi_z + d\phi'_z) & -(d\phi_y + d\phi'_y) \\ -(d\phi_z + d\phi'_z) & 1 & (d\phi_x + d\phi'_x) \\ (d\phi_y + d\phi'_y) & -(d\phi_x + d\phi'_x) & 1 \end{bmatrix} \tag{3-13}$$

$$= \mathbf{R}(d\phi_x + d\phi'_x, d\phi_y + d\phi'_y, d\phi_z + d\phi'_z)$$

where higher order quantities are neglected. Thus, the rotation resulting from two arbitrary infinitesimal rotations is simply given by the algebraic sum of the individual components for each axis, in other words infinitesimal rotations are *additive*. This is another important property of infinitesimal rotations.

Let us write the infinitesimal rotations about the three axes in vector form:

$$d\boldsymbol{\phi} = \begin{bmatrix} d\phi_x \\ d\phi_y \\ d\phi_z \end{bmatrix} \tag{3-14}$$

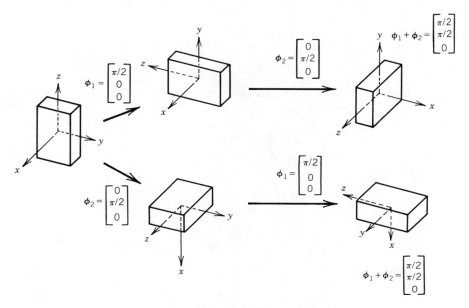

Figure 3-3 : Finite angle rotations.

In general, expressions such as (3-14) can be treated as vectors if they are additive and commutative. As shown above, infinitesimal rotations are additive and commutative. We treat, therefore, the infinitesimal rotations denoted by the expression (3-14) as a vector, because it possesses all the properties that vectors in a vector field must satisfy. Geometrically, the infinitesimal rotation vector $d\phi$ can be represented by an arrow, shown in Figure 3-2. The direction of the arrow represents the axis of rotation, and the length represents the magnitude of the rotation.

It should be noted that vector representation is not allowed for finite rotations, but is valid only for infinitesimal rotations. Figure 3-3 demonstrates that finite rotations cannot be treated as vectors. The rectangular box shown is rotated about the x and y axes in different orders. When we rotate it about the x axis first and then about the y axis, the resultant orientation is given by Figure (a). If the consecutive rotations are carried out in the opposite order, the orientation shown by Figure (b) is obtained, which is completely different from that in Figure (a). However, if we represented the individual rotations by the three-dimensional

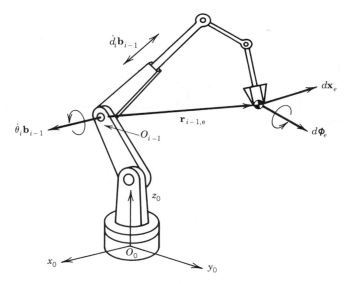

Figure 3-4 :   Infinitesimal displacements of end-effector generated by individual joints.

vectors, $\boldsymbol{\phi}_1$ and $\boldsymbol{\phi}_2$, and used vector addition, the resultant vectors corresponding to the two different orders of rotations would be identical.   Thus finite rotations cannot be treated as vectors.

3.1.3. Computation of the Manipulator Jacobian

In this section we deal with a general n degree-of-freedom manipulator arm and compute the manipulator Jacobian associated with the infinitesimal translation and rotation of the end-effector.   As shown in Figure 3-4, we denote the infinitesimal end-effector translation by a three-dimensional vector $d\mathbf{x}_e$ and we represent the infinitesimal end-effector rotation by a three-dimensional vector $d\boldsymbol{\phi}_e$.   Both vectors are represented with reference to the base coordinate frame $O_0 - x_0 y_0 z_0$.   For convenience, we combine the two vectors and define the following six-dimensional vector $d\mathbf{p}$:

$$d\mathbf{p} = \begin{bmatrix} d\mathbf{x}_e \\ d\boldsymbol{\phi}_e \end{bmatrix} \qquad (3-15)$$

Dividing both sides by the infinitesimal time increment dt, we obtain the velocity and angular velocity of the end-effector:

$$\dot{\mathbf{p}} = \begin{bmatrix} \mathbf{v}_e \\ \omega_e \end{bmatrix} \qquad (3-16)$$

As before, the end-effector velocity and angular velocity can be written as a linear function of joint velocities using the manipulator Jacobian:

$$\dot{\mathbf{p}} = \mathbf{J}\dot{\mathbf{q}} \qquad (3-17)$$

where $\dot{\mathbf{q}} = [\dot{q}_1, \cdots, \dot{q}_n]^T$ is the $n \times 1$ joint velocity vector. The dimension of the Jacobian matrix is now $6 \times n$; the first three row vectors are associated with the linear velocity $\mathbf{v}_e$, while the last three correspond to the angular velocity ω_e. Each column vector, on the other hand, represents the velocity and angular velocity generated by the corresponding individual joint. Let us determine each column vector of the Jacobian matrix as functions of link parameters and arm configuration. Let $\mathbf{J}_{Li}$ and $\mathbf{J}_{Ai}$ be 3×1 column vectors of the Jacobian matrix associated with the linear and angular velocities, respectively. Namely, we partition the Jacobian matrix so that

$$\mathbf{J} = \begin{bmatrix} \mathbf{J}_{L1} \mid \mathbf{J}_{L2} \mid & \mid \mathbf{J}_{Ln} \\ \mid & \mid \cdots \mid \\ \mathbf{J}_{A1} \mid \mathbf{J}_{A2} \mid & \mid \mathbf{J}_{An} \end{bmatrix} \qquad (3-18)$$

Using the $\mathbf{J}_{Li}$, we can write the linear velocity of the end-effector as

$$\mathbf{v}_e = \mathbf{J}_{L1}\dot{q}_1 + \cdots + \mathbf{J}_{Ln}\dot{q}_n \qquad (3-19)$$

If joint i is prismatic, it produces a linear velocity at the end-effector in the same direction as the joint axis. Let $\mathbf{b}_{i-1}$ be the unit vector pointing along the direction of the joint axis i, as shown in Figure 3-4, and let $\dot{d}_i$ be the scalar joint velocity in this direction, then we obtain,

$$\mathbf{J}_{Li}\dot{q}_i = \mathbf{b}_{i-1}\dot{d}_i \qquad (3-20)$$

Note that in the Denavit-Hartenberg notation the joint velocity $\dot{d}_i$ is measured along the z_{i-1}

axis. If the joint is revolute, as shown in the figure, it rotates the composite of distal links from links i to n at the angular velocity $\boldsymbol{\omega}_i$ given by

$$\boldsymbol{\omega}_i = \mathbf{b}_{i-1}\dot{\theta}_i \qquad\qquad (3-21)$$

This angular velocity produces a linear velocity at the end-effector. Let $\mathbf{r}_{i-1,e}$ be the position vector from O_{i-1} to the end-effector as shown in the figure, then the linear velocity generated by the angular velocity $\boldsymbol{\omega}_i$ is given by

$$\mathbf{J}_{Li}\dot{q}_i = \boldsymbol{\omega}_i \times \mathbf{r}_{i-1,e} = (\mathbf{b}_{i-1} \times \mathbf{r}_{i-1,e})\dot{\theta}_i \qquad\qquad (3-22)$$

where $\mathbf{a} \times \mathbf{b}$ represents the vector product of two vectors $\mathbf{a}$ and $\mathbf{b}$. Thus the end-effector velocity is determined by either (3-20) or (3-22) depending on the type of joint.

Similarly, the angular velocity of the end-effector can be expressed as a linear combination of the column vectors $\mathbf{J}_{Ai}$ in equation (3-18),

$$\boldsymbol{\omega}_e = \mathbf{J}_{A1}\dot{q}_1 + \cdots + \mathbf{J}_{An}\dot{q}_n \qquad\qquad (3-23)$$

When joint i is a prismatic joint, it does not generate an angular velocity at the end-effector, hence

$$\mathbf{J}_{Ai}\dot{q}_i = \mathbf{O} \qquad\qquad (3-24)$$

If, on the other hand, the joint is a revolute joint, the angular velocity is given by

$$\mathbf{J}_{Ai}\dot{q}_i = \boldsymbol{\omega}_i = \mathbf{b}_{i-1}\dot{\theta}_i \qquad\qquad (3-25)$$

Equations (3-20), (3-22), (3-24) and (3-25) determine all the elements of the manipulator Jacobian. To summarize:

$$\begin{bmatrix} \mathbf{J}_{Li} \\ \mathbf{J}_{Ai} \end{bmatrix} = \begin{bmatrix} \mathbf{b}_{i-1} \\ \mathbf{O} \end{bmatrix} \qquad\qquad \text{for a prismatic joint} \qquad (3-26)$$

and

$$\begin{bmatrix} \mathbf{J}_{Li} \\ \mathbf{J}_{Ai} \end{bmatrix} = \begin{bmatrix} \mathbf{b}_{i-1} \times \mathbf{r}_{i-1,e} \\ \mathbf{b}_{i-1} \end{bmatrix} \qquad \text{for a revolute joint} \qquad (3-27)$$

Vectors $\mathbf{b}_{i-1}$ and $\mathbf{r}_{i-1,e}$ in the above expressions are functions of joint displacements. These vectors can be computed using the coordinate transformations discussed in the previous chapter. The direction of joint axis $i-1$ is represented by $\overline{\mathbf{b}} = [0,0,1]^T$ with reference to coordinate frame $i-1$, because the joint axis is along the z_{i-1} axis. This vector $\overline{\mathbf{b}}$ can be transformed to a vector which is defined with reference to the base frame, that is $\mathbf{b}_{i-1}$, using 3×3 rotation matrices $\mathbf{R}_j^{j-1}(q_j)$ as:

$$\mathbf{b}_{i-1} = \mathbf{R}_1^0(q_1) \cdots \mathbf{R}_{i-1}^{i-2}(q_{i-1})\overline{\mathbf{b}} \qquad (3-28)$$

Position vector $\mathbf{r}_{i-1,e}$ can be computed by using 4×4 matrices $\mathbf{A}_j^{j-1}(q_j)$. Let $\mathbf{X}_{i-1,e}$ be the 4×1 augmented vector of $\mathbf{r}_{i-1,e}$, and $\overline{\mathbf{X}} = [0,0,0,1]^T$ be the augmented position vector representing the origin of its coordinate frame, then the position vector $\mathbf{r}_{i-1,e}$ is derived from

$$\mathbf{X}_{i-1,e} = \mathbf{A}_1^0(q_1) \cdots \mathbf{A}_n^{n-1}(q_n)\overline{\mathbf{X}} - \mathbf{A}_1^0(q_1) \cdots \mathbf{A}_{i-1}^{i-2}(q_{i-1})\overline{\mathbf{X}} \qquad (3-29)$$

where the first term accounts for the position vector from the origin O_0 to the end-effector and the second term is from O_0 to O_{i-1}.

Example 3-1: Three Degree-of-Freedom Polar Coordinate Manipulator

Let us work out an example of the manipulator Jacobian computation. The skeleton structure of a 3 degree-of-freedom manipulator is shown in Figure 3-5. The joint displacements θ_1, θ_2, and d_3 defined in the figure are equivalent to polar coordinates; hence this manipulator arm is called a polar coordinate manipulator. To find the Jacobian matrix, we begin by determining the directions of the joint axes. From the figure these are given by

$$\mathbf{b}_0 = \begin{bmatrix} 0 \\ 0 \\ 1 \end{bmatrix} \qquad \mathbf{b}_1 = \begin{bmatrix} -s_1 \\ c_1 \\ 0 \end{bmatrix} \qquad \mathbf{b}_2 = \begin{bmatrix} c_1 s_2 \\ s_1 s_2 \\ c_2 \end{bmatrix} \qquad (3-30)$$

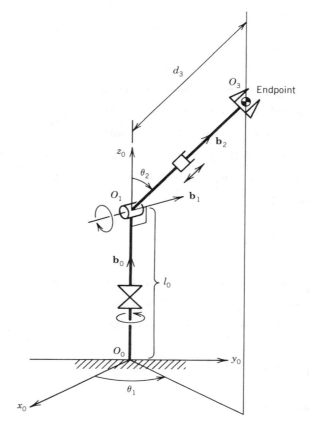

Figure 3-5 :   Three degree-of-freedom spherical coordinate manipulator.

For revolute joints, we need to find position vector $\mathbf{r}_{i-1,e}$. They are

$$\mathbf{r}_{1,e} = d_3\mathbf{b}_2$$

$$\mathbf{r}_{0,e} = l_0\mathbf{b}_0 + d_3\mathbf{b}_2$$

(3–31)

Substituting (3-30) and (3-31) into (3-18), (3-26) and (3-27),

$$\mathbf{J} = \begin{bmatrix} -d_3 s_1 s_2 & d_3 c_1 c_2 & c_1 s_2 \\ d_3 c_1 s_2 & d_3 s_1 c_2 & s_1 s_2 \\ 0 & -d_3 s_2 & c_2 \\ 0 & -s_1 & 0 \\ 0 & c_1 & 0 \\ 1 & 0 & 0 \end{bmatrix}$$

(3–32)

Note that the elements of the manipulator Jacobian are functions of the joint displacements, hence the Jacobian is configuration-dependent. $\Delta\Delta\Delta$

3.2. Inverse Instantaneous Kinematics

3.2.1. Resolved Motion Rate

Equation (3-17) in the previous section provided the velocity and angular velocity of the end-effector as a linear function of joint velocities. Using this expression we now discuss the inverse problem. Namely, given a desired end-effector velocity, we find the corresponding joint velocities that cause the end-effector to move at the desired velocity.

As mentioned in Chapter 2, a manipulator arm must have at least six degrees of freedom in order to locate its end-effector at an arbitrary position with an arbitrary orientation. Similarly, six degrees of freedom are also necessary to move the end-effector in an arbitrary direction with an arbitrary angular velocity. In this section, we discuss the inverse problem for six degree-of-freedom manipulators. In Section 3.2.2, we will extend the derivation to general n degree-of-freedom manipulators.

For a six degree-of-freedom manipulator, the Jacobian matrix $\mathbf{J}$ is a 6×6 square matrix. If the matrix is non-singular at the current arm configuration, the inverse matrix $\mathbf{J}^{-1}$ exists. We then obtain from (3-17)

$$\dot{\mathbf{q}} = \mathbf{J}^{-1}\dot{\mathbf{p}} \qquad\qquad (3\text{--}33)$$

Equation (3-33) determines the velocities required at the individual joints in order to obtain a given end-effector velocity $\dot{\mathbf{p}}$ — this scheme is referred to as *resolved motion rate control* and is attributed to (Whitney, 1969). Since the manipulator Jacobian varies with the arm configuration, it may become singular at certain configurations. In such cases the inverse Jacobian does not exist, hence solution (3-33) is not valid. The corresponding arm configuration is then itself called *singular*. At a singular configuration, the matrix $\mathbf{J}$ is not of full rank; hence, its column vectors are linearly dependent, and thus do not span the whole six-dimensional vector space of $\dot{\mathbf{p}}$. Therefore, there exists at least one direction in which the end-

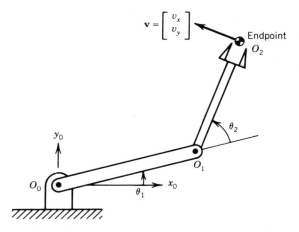

Figure 3-6 :   Endpoint velocities of the two d.o.f. planar manipulator.

effector cannot be moved no matter how we choose joint velocities $\dot{q}_1$ through $\dot{q}_6$. Let us work out an example of this effect.

Example 3-2

Consider the two degree-of-freedom planar manipulator shown in Figure 3-6.   The length of each link is 1, and the endpoint velocity is denoted by $\mathbf{v} = [v_x, v_y]^T$.

1. Given a desired endpoint velocity, find joint velocities that produce the desired endpoint velocity.

2. Find singular configurations, and determine in which direction the endpoint cannot move for each singular configuration.

3. Find profiles of joint velocities when the endpoint is required to track the trajectory shown in Figure 3-7 at a constant tangential speed.

1.   The Jacobian matrix for this planar manipulator has been derived in equation (3-6). Replacing l_1 and l_2 by 1, we obtain

$$\mathbf{J} = \begin{bmatrix} -\sin \theta_1 - \sin (\theta_1 + \theta_2), & -\sin (\theta_1 + \theta_2) \\ \cos \theta_1 + \cos (\theta_1 + \theta_2), & \cos (\theta_1 + \theta_2) \end{bmatrix} \qquad (3\text{--}34)$$

Inverting the Jacobian matrix and substituting into (3-33), the joint velocities are given by

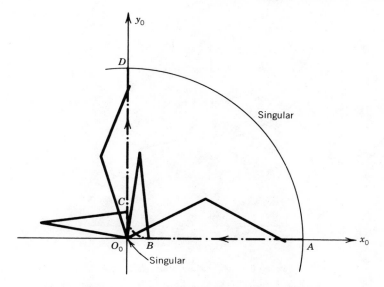

Figure 3-7 : Trajectory tracking near singular points.

$$\dot{\theta}_1 = \frac{v_x\cos{(\theta_1+\theta_2)} + v_y\sin{(\theta_1+\theta_2)}}{\sin{\theta_2}} \qquad (3\text{--}35)$$

$$\dot{\theta}_2 = -\frac{v_x[\cos{\theta_1}+\cos{(\theta_1 + \theta_2)}] + v_y[\sin{\theta_1}+\sin{(\theta_1 + \theta_2)}]}{\sin{\theta_2}} \qquad (3\text{--}36)$$

2.   Singularity occurs when the determinant of the manipulator Jacobian is zero. Now from expression (3-34)

$$\det(\mathbf{J}) = \sin{\theta_2} \qquad (3\text{--}37)$$

Therefore, singular configurations occur for $\theta_2 = 0$ or $\theta_2 = \pi$, i.e. when the arm is fully extended or fully contracted. This corresponds to the endpoint positions shown in Figure 3-7, so that the origin O_0 and the boundary of the reachable space are singular. At the singular configuration, equation (3-19) becomes

$$\begin{bmatrix} v_x \\ v_y \end{bmatrix} = \begin{bmatrix} -2\sin{(\theta_1)} \\ 2\cos{(\theta_1)} \end{bmatrix} \dot{\theta}_1 + \begin{bmatrix} -\sin{(\theta_1)} \\ \cos{(\theta_1)} \end{bmatrix} \dot{\theta}_2$$

$$= \begin{bmatrix} -\sin{(\theta_1)} \\ \cos{(\theta_1)} \end{bmatrix} (2\dot{\theta}_1+\dot{\theta}_2) \qquad (3\text{--}38)$$

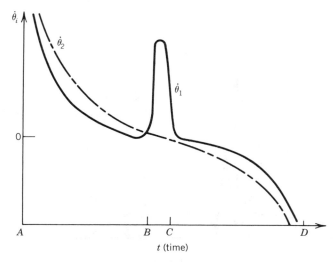

Figure 3-8 : Profiles of joint velocities.

i.e., the two column vectors of the Jacobian matrix become parallel. The endpoint can then be moved only in the direction perpendicular to the arm links, but not in any other direction.

3.   To find the velocity profile for tracking the specified trajectory, we first obtain the joint angles that correspond to each endpoint position on the trajectory. Then we substitute the joint angles into (3-35) and (3-36), and determine the joint velocities required. The result is shown in Figure 3-8, where the two joint velocities are plotted with respect to time. Note that both velocities are excessively large near the singular points, A and D. To generate the endpoint velocity in the radial directions, OA and OD, excessively large velocities are required for both joints. In this region, the denominators in equations (3-35) and (3-36) are almost zero. Also, the velocity of the first joint becomes excessively large between points B and C, since the arm links must rotate quickly in this region. Again, this region is near the singular point. Thus, even if the inverse of the manipulator Jacobian exists, the joint velocities become excessively large in the vicinity of singular points.   ▲▲▲

3.2.2. Redundancy

We have seen that there are a finite number of solutions to the kinematic equation of a six degree-of-freedom manipulator arm. For the instantaneous kinematic equation, a unique solution exists if the arm configuration is non-singular. However, when a manipulator arm has more than six degrees of freedom, we can find an infinite number of solutions that provide the same motion required for the end-effector. Consider for instance the human arm, which has seven degrees of freedom excluding the joints at the fingers. As seen in Section 2.3.1, when the hand is placed on a desk and fixed in its position and orientation, the elbow position can still vary continuously without moving the hand. This implies that a certain ratio of joint velocities exists that does not cause any velocity at the hand. This specific ratio of joint velocities does not contribute to the resultant endpoint velocity. Even if these joint velocities are superimposed to other joint velocities, the resultant end-effector velocity is the same. Consequently, we can find different solutions of the instantaneous kinematic equation for the same end-effector velocity. In this section, we investigate the fundamental properties of the instantaneous kinematics when additional degrees of freedom are involved.

To formulate the instantaneous kinematic equation for a general n degrees-of-freedom manipulator arm, we begin by modifying the definition of the vector $d\mathbf{p}$ representing the end-effector motion. In equation (3-15), $d\mathbf{p}$ was defined as a six-dimensional vector that represents the infinitesimal translation and rotation of an end-effector. However, the number of variables that we need to specify for performing a task is not necessarily six. In arc welding, for example, only five independent variables of torch motion need be controlled. Since the welding torch is usually symmetric about its center line, we can locate the torch with an arbitrary orientation about the center line. Thus five degrees of freedom are sufficient to perform arc welding operations. In general, we describe the end-effector motion by m independent variables that must be specified to perform a given task. Let $d\mathbf{p} = [dp_1, \cdots, dp_m]^T$ be the $m \times 1$ vector which represents the infinitesimal displacements of the end-effector, then the instantaneous kinematic equation for a general n degree-of-freedom manipulator arm is given by

$$d\mathbf{p} = \mathbf{J}d\mathbf{q} \qquad\qquad (3-39)$$

where the dimension of the manipulator Jacobian $\mathbf{J}$ is $m \times n$. When n is larger than m and $\mathbf{J}$ is of full rank, there are $(n\text{-}m)$ arbitrary variables in the general solution of (3-39). The manipulator is then said to have $(n\text{-}m)$ *redundant degrees of freedom* for the given task.

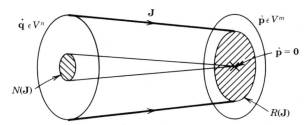

Figure 3-9 :   Linear mapping diagram of instantaneous kinematics.

The Jacobian matrix also determines the relationship between the end-effector velocity $\dot{\mathbf{p}}$ and joint velocities $\dot{\mathbf{q}}$:

$$\dot{\mathbf{p}} = \mathbf{J}\dot{\mathbf{q}} \tag{3-40}$$

Equation (3-40) can be regarded as a linear mapping from n-dimensional vector space $\mathbf{V}^n$ to m-dimensional space $\mathbf{V}^m$. To characterize the solution to equation (3-40), let us interpret the equation using the linear mapping diagram shown in Figure 3-9. The subspace $R(\mathbf{J})$ in the figure is the range space of the linear mapping. The range space represents all the possible end-effector velocities that can be generated by the n joints at the present arm configuration. If the rank of $\mathbf{J}$ is of full row rank, the range space covers the entire vector space $\mathbf{V}^m$. Otherwise, there exists at least one direction in which the end-effector cannot be moved. The subspace $N(\mathbf{J})$ of Figure 3-9 is the null space of the linear mapping. Any element in this subspace is mapped into the zero vector in $\mathbf{V}^m$: $\mathbf{J}\dot{\mathbf{q}} = \mathbf{0}$. Therefore, any joint velocity vector $\dot{\mathbf{q}}$ that belongs to the null space does not produce any velocity at the end-effector. Recall the human arm discussed before. The elbow can move without moving the hand. Joint velocities for this motion are involved in the null space, since no end-effector motion is induced. If the manipulator Jacobian is of full rank, the dimension of the null space, dim $N(\mathbf{J})$, is the same as the number of redundant degrees of freedom (n-m). When the Jacobian matrix is degenerate, i.e. not of full rank, the dimension of the range space, dim $R(\mathbf{J})$, decreases, and at the same time the dimension of the null space increases by the same amount. The sum of the two is always equal to n:

$$\dim R(\mathbf{J}) + \dim N(\mathbf{J}) = n \tag{3-41}$$

Let $\dot{\mathbf{q}}^{*}$ be a solution of (3-40) and $\dot{\mathbf{q}}_0$ be a vector involved in the null space, then the vector of the form $\dot{\mathbf{q}} = \dot{\mathbf{q}}^{*} + k\dot{\mathbf{q}}_0$ is also a solution of (3-40), where k is an arbitrary scalar quantity. Namely,

$$\mathbf{J}\dot{\mathbf{q}} = \mathbf{J}\dot{\mathbf{q}}^{*} + k\mathbf{J}\dot{\mathbf{q}}_0 = \mathbf{J}\dot{\mathbf{q}}^{*} = \dot{\mathbf{p}} \tag{3-42}$$

Since the second term $k\dot{\mathbf{q}}_0$ can be chosen arbitrarily within the null space, an infinite number of solutions exists for the linear equation, unless the dimension of the null space is 0. The null space accounts for the arbitrariness of the solutions. The general solution to the linear equation involves the same number of arbitrary parameters as the dimension of the null space.

3.2.3. Optimal Solutions

In this section we discuss an optimal solution to the velocity relationship (3-40). We fix the manipulator Jacobian at an appropriate arm configuration, and find the optimal solution to the linear equation (3-40), assuming that the Jacobian matrix is of full row rank. We evaluate solutions to the linear equation by the quadratic cost function of the joint velocity vector given by

$$G(\dot{\mathbf{q}}) = \dot{\mathbf{q}}^{T}\mathbf{W}\dot{\mathbf{q}} \tag{3-43}$$

where $\mathbf{W}$ is an $n \times n$ symmetric positive definite weighting matrix. The problem is to find the $\dot{\mathbf{q}}$ that satisfies equation (3-40) for a given $\dot{\mathbf{p}}$ and $\mathbf{J}$ while minimizing the cost function $G(\dot{\mathbf{q}})$. Let us solve this problem using Lagrange multipliers. To this end we use a modified cost function of the form

$$G(\dot{\mathbf{q}},\boldsymbol{\lambda}) = \dot{\mathbf{q}}^{T}\mathbf{W}\dot{\mathbf{q}} - \boldsymbol{\lambda}^{T}(\mathbf{J}\dot{\mathbf{q}} - \dot{\mathbf{p}}) \tag{3-44}$$

where $\boldsymbol{\lambda}$ is an $m \times 1$ unknown vector of Lagrange multipliers. The necessary conditions that the optimal solution must satisfy are

$$\frac{\partial G}{\partial \dot{\mathbf{q}}} = 0 \quad , \quad \text{that is,} \quad 2\mathbf{W}\dot{\mathbf{q}} - \mathbf{J}^{T}\boldsymbol{\lambda} = 0 \tag{3-45}$$

and

$$\frac{\partial G}{\partial \boldsymbol{\lambda}} = 0 \quad , \quad \text{that is,} \quad \mathbf{J}\dot{\mathbf{q}} - \dot{\mathbf{p}} = 0 \qquad (3\text{--}46)$$

which is of course identical to (3-40). Now matrix $\mathbf{W}$ is positive definite, hence invertible. Thus, we obtain from equation (3-45)

$$\dot{\mathbf{q}} = \tfrac{1}{2}\, \mathbf{W}^{-1} \mathbf{J}^T \boldsymbol{\lambda} \qquad (3\text{--}47)$$

Substituting the above into (3-46) yields

$$(\mathbf{J}\mathbf{W}^{-1}\mathbf{J}^T)\boldsymbol{\lambda} = 2\dot{\mathbf{p}} \qquad (3\text{--}48)$$

Since J is assumed to be of full row-rank, matrix product $\mathbf{J}\mathbf{W}^{-1}\mathbf{J}^T$ is a full-rank square matrix, and is therefore invertible. Eliminating the Lagrange multiplier vector $\boldsymbol{\lambda}$ in equations (3-47) and (3-48), we obtain the optimal solution

$$\dot{\mathbf{q}} = \mathbf{W}^{-1}\mathbf{J}^T(\mathbf{J}\mathbf{W}^{-1}\mathbf{J}^T)^{-1}\dot{\mathbf{p}} \qquad (3\text{--}49)$$

Clearly, the above solution satisfies the original velocity relationship (3-40). Indeed, we can obtain equation (3-40) by premultiplying equation (3-49) by the Jacobian matrix J. When the weighting matrix $\mathbf{W}$ is the $m \times m$ identity matrix, the above solution reduces to

$$\dot{\mathbf{q}} = \mathbf{J}^T(\mathbf{J}\mathbf{J}^T)^{-1}\dot{\mathbf{p}} \qquad (3\text{--}50)$$

The matrix product $\mathbf{J}^{\#} = \mathbf{J}^T(\mathbf{J}\mathbf{J}^T)^{-1}$ is known as the *pseudo-inverse* of the Jacobian matrix.

3.3. Research Topics

The computation of the manipulator Jacobian is time-intensive, which is a crucial problem for real-time control. Efficient computation algorithms are suggested by (Orin and Schrader, 1984; Pennock and Yang, 1985-b). (Sugimoto,1984) studies the derivation of joint velocities from endpoint velocities without explicit computation of the Jacobian.

Singularity is a critical problem for articulated manipulator arms. (Waldron, Wang and Bolin, 1985; Litvin and Castelli, 1985) analyze arm singularities, while (Paul and Stevenson,

1983; Asada and Cro Granito, 1985) address wrist singularities and methods for avoiding them. (Hollerbach, 1984) discusses the kinematic structures of redundant manipulators that are appropriate for avoiding the singular points internal to the workspace.

The study of redundant manipulators is an important research topic in advanced manipulation, particularly for obstacle and singularity avoidance . The application of pseudo-inverse matrices to obtain optimal joint velocities has traditionally been a central tool for the redundant arm problem (Hanafusa, *et al.*, 1981; Konstantinov, *et al.*, 1982; Klein, 1983; Nakamura and Hanafusa, 1984; Hollerbach, 1984-b).

Resolved motion rate control has been extended to resolved acceleration control by several researchers. Computational efficiency is again a central theme (Luh, Walker and Paul, 1980-b; Hollerbach, 1983).

The kinematic analysis and design problems have been extended to closed-loop mechanisms, in particular mechanical fingers (Hanafusa and Asada, 1976; Asada, 1979; Salisbury, 1982 and 1984; Salisbury and Craig, 1982; Okada, 1979; Kobayashi, 1984; Bakar *et al.*, 1985; Holzmann and McCarthy, 1985) The kinematic problems associated with the manipulation of rigid bodies with mechanical fingers has been studied in the contexts of automatic assembly (Ohwovoriole and Roth, 1981) and workpart fixturing (Asada and By, 1984, 1985).

Chapter 4
STATICS

The contact between a manipulator and its environment results in interactive forces and moments at the manipulator/environment interface. In this chapter we focus upon the forces and moments which act on a manipulator arm when it is at rest.

Each joint of a manipulator arm is driven by an individual actuator. The corresponding input joint torques are transmitted through the arm linkage to the end-effector, where the resultant force and moment act upon the environment. The relationship between the actuator drive torques and the resultant force and moment applied at the manipulator endpoint is one of the major issues discussed in this chapter. This input-output relationship is of fundamental importance in the control of a manipulator arm.

The relationship between the force and moment applied by the environment and the resultant deflection of the arm linkage is also discussed in this chapter. If a manipulator is used to carry a heavy object at its endpoint or if a large force is applied to the end-effector, the endpoint of the manipulator arm deflects. The magnitude of this deflection is directly determined by the stiffness of the manipulator arm. Endpoint stiffness is an important characteristic that determines the strength and accuracy of the manipulator. It also plays an important role in the control of mechanical interactions with the environment, as discussed later in Chapter 7.

4.1. Force and Moment Analysis

4.1.1. Balance of Forces and Moments

In this section, we derive the basic equations that govern the static behavior of a manipulator arm.

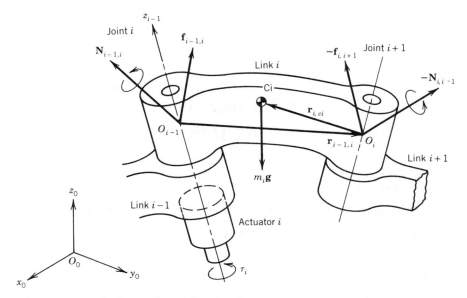

Figure 4-1 :   Forces and moments acting on link i.

We begin by considering the free body diagram of an individual link incorporated in an open kinematic chain. Figure 4-1 shows the forces and moments acting on link i, which is connected to link $i-1$ and link $i+1$ by joint i and joint $i+1$, respectively. The linear force acting at point 0_{i-1} , that is the origin of the coordinate frame $0_{i-1}-x_{i-1}y_{i-1}z_{i-1}$, is denoted by vector $\mathbf{f}_{i-1,i}$, where the force is exerted by the link of the first subscript and acts upon the link of the second subscript. The vector $\mathbf{f}_{i,i+1}$, therefore, represents the force applied to link $i+1$ by link i. The force applied to link i by link $i+1$ is then given by $-\mathbf{f}_{i,i+1}$. The gravity force acting at the centroid C_i is denoted by $m_i\mathbf{g}$, where m_i is the mass of link i and $\mathbf{g}$ is the 3×1 vector representing the acceleration of gravity. The balance of linear forces is then given by

$$\mathbf{f}_{i-1,i} - \mathbf{f}_{i,i+1} + m_i\mathbf{g} = \mathbf{0}, \qquad i=1, \cdots ,n \tag{4-1}$$

Note that all the vectors are defined with reference to the base coordinate frame $0_0-x_0y_0z_0$.

Next, we derive the balance of moments. The moment applied to link i by link $i-1$ is denoted by $\mathbf{N}_{i-1,i}$, and therefore the moment applied to link i by link $i+1$ is $-\mathbf{N}_{i,i+1}$

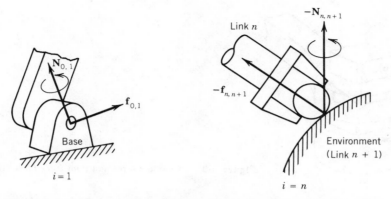

Figure 4-2 :   Forces and moments exerted by the base link and the environment.

Further, the linear forces $\mathbf{f}_{i-1,i}$ and $-\mathbf{f}_{i,i+1}$ also cause moments about the centroid C_i' . The balance of moments with respect to the centroid C_i is thus given by

$$\mathbf{N}_{i-1,i} - \mathbf{N}_{i,i+1} - (\mathbf{r}_{i-1,i} + \mathbf{r}_{i,ci}) \times \mathbf{f}_{i-1,i} + (-\mathbf{r}_{i,ci}) \times (-\mathbf{f}_{i,i+1}) = 0, \quad i = 1, \cdots, n \ (4\text{-}2)$$

where $\mathbf{r}_{i-1,i}$ is the 3×1 position vector from the point 0_{i-1} to point 0_i with reference to the base coordinate frame, and $\mathbf{r}_{i,ci}$ represents the position vector from the point 0_i to the centroid C_i . The force $\mathbf{f}_{i-1,i}$ and moment $\mathbf{N}_{i-1,i}$ are called the *coupling force and moment* between the adjacent links i and $i-1$. When $i = 1$, the coupling force and moment becomes $\mathbf{f}_{0,1}$ and $\mathbf{N}_{0,1}$. These are interpreted as the reaction force and moment applied to the base link to which the arm linkage is fixed (see Figure 4-2). When $i = n$, on the other hand, the above coupling force and moment become $\mathbf{f}_{n,n+1}$ and $\mathbf{N}_{n,n+1}$. As shown in Figure 4-2, when the end-effector (that is, link n) contacts the environment, the reaction force and moment act on the last link. For convenience we regard the environment as an additional link, numbered $n+1$, and we represent the reaction force and moment by $-\mathbf{f}_{n,n+1}$ and $-\mathbf{N}_{n,n+1}$, respectively.

The above two equations can be derived for all the link members except the base link, $i = 1, \cdots, n$. The total number of vector equations that we can derive is then $2n$, whereas the number of coupling forces and moments involved is $2(n+1)$. Therefore two of the coupling forces and moments must be specified; otherwise the equations cannot be solved. The final coupling force and moment, $\mathbf{f}_{n,n+1}$ and $\mathbf{N}_{n,n+1}$, are the force and moment that the

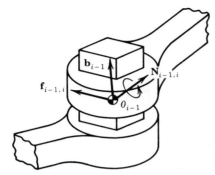

Figure 4-3 :   Coupling force and joint force for a prismatic joint.

manipulator arm applies to the environment. To perform a task successfully, the manipulator arm needs to accommodate this force and moment. Thus, we specify this coupling force and moment, and solve the simultaneous equations. For convenience we combine the force $\mathbf{f}_{n,n+1}$ and the moment $\mathbf{N}_{n,n+1}$ and define the following six-dimensional vector,

$$\mathbf{F} = \begin{bmatrix} \mathbf{f}_{n,n+1} \\ \mathbf{N}_{n,n+1} \end{bmatrix} \tag{4-3}$$

We call $\mathbf{F}$ the *endpoint force and moment vector* or simply the *endpoint force.*

4.1.2. Equivalent Joint Torques

In this section we derive the functional relationship between the input torques exerted by the actuators and the resultant endpoint force. We assume that each joint is driven by an individual actuator that exerts a drive torque or force between the adjacent links. Let τ_i be the drive torque or force exerted by the i-th actuator driving joint i.

For a prismatic joint, the drive force τ_i is a linear force exerted along the joint axis $i-1$, as shown in Figure 4-3. Assuming that the joint mechanism is frictionless, we can relate the joint force τ_i to the linear coupling force $\mathbf{f}_{i-1,i}$ between links $i-1$ and i by

$$\tau_i = \mathbf{b}_{i-1}^T \cdot \mathbf{f}_{i-1,i} \tag{4-4}$$

where $\mathbf{b}_{i-1}$ represents the unit vector pointing in the direction of the joint axis and $\mathbf{a}^T \cdot \mathbf{b}$ represents the inner product of vectors $\mathbf{a}$ and $\mathbf{b}$. Equation (4-4) implies that the actuator must bear only the component of $\mathbf{f}_{i-1,i}$ which is in the direction of the joint axis, and that the components in all the other directions are supported by the joint structure. These components of the coupling force are internal constraint forces, which do not produce work.

For a revolute joint, τ_i represents a drive torque rather than a linear force. This drive torque is balanced with the coupling torque component of $\mathbf{N}_{i-1,i}$ which is in the direction of its joint axis:

$$\tau_i = \mathbf{b}_{i-1}^T \cdot \mathbf{N}_{i-1,i} \tag{4-5}$$

Other components of the coupling torque $\mathbf{N}_{i-1,i}$ are borne by the joint structure. They are workless constraint moments.

We can combine all the joint forces and torques together to define the n-vector given by

$$\boldsymbol{\tau} = \begin{bmatrix} \tau_1 \\ \tau_2 \\ \vdots \\ \tau_n \end{bmatrix} \tag{4-6}$$

We call $\boldsymbol{\tau}$ the *joint torque and force vector* or simply the *joint torques*. The joint torques represent the actuators' inputs to the arm linkage. The relationship between the joint torques $\boldsymbol{\tau}$ and the endpoint force vector $\mathbf{F}$ is stated by the following theorem:

Theorem

Assume that the joint mechanisms are frictionless, then the joint torques $\boldsymbol{\tau}$ that are required to bear an arbitrary endpoint force $\mathbf{F}$ are given by

$$\boldsymbol{\tau} = \mathbf{J}^T \mathbf{F} \tag{4-7}$$

where $\mathbf{J}$ is the $6 \times n$ manipulator Jacobian relating infinitesimal joint displacements $d\mathbf{q}$ to infinitesimal end-effector displacements $d\mathbf{p}$:

$$d\mathbf{p} = \mathbf{J} d\mathbf{q} \tag{4-8}$$

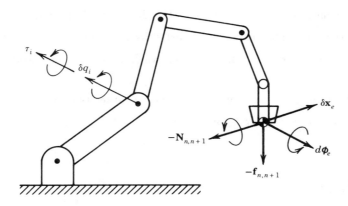

Figure 4-4 :   Virtual displacements of end effector and individual joints.

Note that the above joint torques do not account for gravity torques or any other torques. They are the net torques that balance the *endpoint* force and moment. We call τ of equation (4-7) the *equivalent joint torques* corresponding to the endpoint force $\mathbf{F}$.

Proof

We prove the theorem using the *principle of virtual work*. Consider *virtual displacements* at individual joints, δq_i, and at the end-effector, $\delta \mathbf{x}_e$ and $\delta \boldsymbol{\phi}_e$, as shown in Figure 4-4. Virtual displacements are arbitrary infinitesimal displacements of a mechanical system that conform to the geometric constraints of the system. Virtual displacements are different from actual displacements, in that they must only satisfy *geometric* constraints and do not have to meet other laws of motion. To distinguish the virtual displacements from the actual displacements, we use the greek letter δ rather than the roman d. We assume that joint torques $\tau_i (i=1, \cdots, n)$ and endpoint force and moment, $-\mathbf{f}_{n,n+1}$ and $-\mathbf{N}_{n,n+1}$, act on the manipulator while the joints and the end-effector are displaced. Then, the virtual work done by the forces and moments is given by

$$\delta Work = \tau_1 \delta q_1 + \cdots + \tau_n \delta q_n - \mathbf{f}^T_{n,n+1} \, \delta \mathbf{x}_e - \mathbf{N}^T_{n,n+1} \, \delta \boldsymbol{\phi}_e$$

or

$$\delta Work = \boldsymbol{\tau}^T \delta \mathbf{q} - \mathbf{F}^T \delta \mathbf{p} \tag{4-9}$$

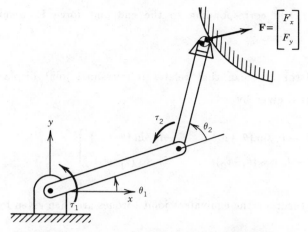

Figure 4-5 :   Endpoint force and equivalent joint torques.

According to the principle of virtual work, the arm linkage is in equilibrium if, and only if, the virtual work *δWork* vanishes for arbitrary virtual displacements, which conform to geometric constraints. Note that the virtual displacements $\delta\mathbf{q}$ and $\delta\mathbf{p}$ are not independent but are related by the manipulator Jacobian to conform to the geometric constraint imposed by the arm linkage. Using (4-8) we can rewrite (4-9) as

$$\delta Work = \boldsymbol{\tau}^T\delta\mathbf{q} - \mathbf{F}^T\mathbf{J}\delta\mathbf{q} = (\boldsymbol{\tau} - \mathbf{J}^T\mathbf{F})^T\delta\mathbf{q} \qquad\qquad (4-10)$$

The above expression only involves $\delta\mathbf{q}$, which represents independent variables for geometrically admissible displacements. In order for (4-10) to vanish for arbitrary $\delta\mathbf{q}$, we must have:

$$\boldsymbol{\tau} = \mathbf{J}^T\,\mathbf{F}$$

i.e., equation (4-7).   △△△

Example 4-1

    Figure 4-5 shows a two degree-of-freedom planar manipulator. At the endpoint, the arm is in contact with the external surface and applies the force $\mathbf{F}=[F_x,F_y]^T$. Find the equivalent joint torques $\boldsymbol{\tau} = [\tau_1, \tau_2]^T$ corresponding to the endpoint force $\mathbf{F}$, assuming that the joint mechanisms are frictionless.

The manipulator Jacobian that relates infinitesimal joint displacements to the end-effector displacement is given by

$$\mathbf{J} = \begin{bmatrix} -l_1 \sin \theta_1 - l_2 \sin (\theta_1 + \theta_2) & -l_2 \sin (\theta_1 + \theta_2) \\ l_1 \cos \theta_1 + l_2 \cos (\theta_1 + \theta_2) & l_2 \cos (\theta_1 + \theta_2) \end{bmatrix} \qquad (4-11)$$

From the preceding theorem, the equivalent joint torques are then given by

$$\begin{bmatrix} \tau_1 \\ \tau_2 \end{bmatrix} = \begin{bmatrix} -l_1 \sin \theta_1 - l_2 \sin (\theta_1 + \theta_2) & l_1 \cos \theta_1 + l_2 \cos (\theta_1 + \theta_2) \\ -l_2 \sin (\theta_1 + \theta_2) & l_2 \cos (\theta_1 + \theta_2) \end{bmatrix} \begin{bmatrix} F_x \\ F_y \end{bmatrix} \qquad (4-12)$$

◭◭◭

4.1.3. Duality

We have found that the equivalent joint torques are related to the endpoint force by the manipulator Jacobian, which is the same matrix that relates the infinitesimal joint displacements to the end-effector displacement. Thus, the static force relationship is closely related to the instantaneous kinematics. In this section we discuss the physical meaning of this observation.

To interpret the similarity between kinematics and statics, we can use the linear mapping diagram of Figure 3-9. Recall that the instantaneous kinematic equation can be regarded as a linear mapping when the Jacobian matrix is fixed at a given arm configuration. Figure 4-6 reproduces Figure 3-9 and completes it with a similar diagram corresponding to the static analysis. As before, the range space $R(\mathbf{J})$ represents the set of all the possible end-effector velocities generated by joint motions. When the Jacobian matrix is degenerate or the arm configuration is singular, the range space does not span the whole vector space V^m. Namely, there exists a direction in which the end-effector cannot move. The null space $N(\mathbf{J})$, on the other hand, represents the set of joint velocities that do not produce a velocity at the

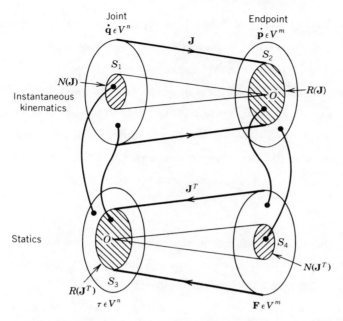

Figure 4-6 :   Linear mapping diagrams for statics and instantaneous kinematics.

end-effector. If the null space is not an empty set, the instantaneous kinematic equation has an infinite number of solutions that cause the same end-effector velocity.

Another linear mapping is associated with the static force relationship (4-7), as shown in the figure. Unlike the mapping of instantaneous kinematics, the mapping of static forces is from the m-dimensional vector space V^m, associated with the end-effector coordinates, to the n-dimensional vector space V^n, associated with the joint coordinates. Therefore the joint torques τ are always determined uniquely for any arbitrary endpoint force $\mathbf{F}$. However, for given joint torques, a balancing endpoint force does not always exist. As in the case of the instantaneous kinematics, let us define the null space $N(\mathbf{J}^T)$ and the range space $R(\mathbf{J}^T)$ of the static force mapping. The null space $N(\mathbf{J}^T)$ represents the set of all endpoint forces that do not require any torques at the joints to bear the corresponding load. In this case the endpoint force is borne entirely by the structure of the arm linkage. The range space $R(\mathbf{J}^T)$, on the other hand, represents the set of all the possible joint torques that can balance the endpoint forces.

The ranges and null spaces of $\mathbf{J}$ and $\mathbf{J}^T$ are closely related. According to the rules of linear algebra, the null space $N(\mathbf{J})$ is the orthogonal complement of the range space $R(\mathbf{J}^T)$. Namely, if a non-zero n-vector $\mathbf{x}$ is in $N(\mathbf{J})$, it cannot also belong to $R(\mathbf{J}^T)$, and vice-versa. If we denote by S_1 the orthogonal complement of $N(\mathbf{J})$, then the range space $R(\mathbf{J}^T)$ is identical to S_1, as shown in the figure. Also, space S_3, i.e., the orthogonal complement of $R(\mathbf{J}^T)$, is identical to $N(\mathbf{J})$. What this implies is that, in the direction in which joint velocities do not cause any end-effector velocity, the joint torques cannot be balanced with any endpoint force. In order to maintain a stationary arm configuration, the joint torques in this space must be zero.

There is a similar correspondence in the end-effector coordinate space V^m. The range space $R(\mathbf{J})$ is the orthogonal complement to the null space $N(\mathbf{J}^T)$. Hence, the space S_2 in the figure is identical to $N(\mathbf{J}^T)$, and the space S_4 is identical to $R(\mathbf{J})$. Therefore, no joint torques are required to balance the end point force when the external force acts in the direction in which the end-effector cannot be moved by the motion of the arm joints. Also, when the external endpoint force is applied in the direction along which the end-effector can move, the external force must be borne entirely by the joint torques. When the Jacobian matrix is degenerate or the arm is in a singular configuration, the null space $N(\mathbf{J}^T)$ is not reduced to $\mathbf{0}$, and the external force can be borne by the arm structure in part. Thus, instantaneous kinematics and statics are closely related. This relationship is referred to as the *duality* of kinematics and statics.

4.1.4. Transformations of Forces and Moments

In the previous section we found that static forces and moments can be analyzed as an instantaneous kinematics problem, using the duality law in the instantaneous kinematics and statics. Once we set up the instantaneous kinematic equation, we can derive the relationship between static forces and moments immediately without going through free dody diagrams. This force and moment analysis utilizing the duality law can be extended to other problems that we encounter in the design and control of manipulator arms. In this section, we generalize the duality law and apply it to a robotics problem.

We begin by modifying the definition of vector $\mathbf{q}$, which was defined as joint displacements. Let $\mathbf{q}=[q_1, \cdots ,q_n]^T$ be defined instead as any independent set of generalized coordinates that locate a mechanical system completely. Joint displacements are an instance of such independent and complete sets of generalized coordinates. Let $\mathbf{Q}=[Q_1, \cdots ,Q_n]^T$ be the generalized forces corresponding to the generalized coordinates $\mathbf{q}$. We also assume that there exists another set of generalized coordinates denoted by $\mathbf{p}=[p_1, \cdots ,p_m]^T$. Note that the $\mathbf{p}$ coordinates do not have to be complete, namely, all the degrees of freedom of the system are not necessarily determined by the set of the coordinates. For example, the position and orientation of an end-effector does not determine the whole configuration of a manipulator arm when the manipulator arm has redundant degrees of freedom. In this case an infinite number of configurations correspond to the same end-effector location. We consider an instant when static forces and moments act upon the system located at $\mathbf{q}$. Suppose that the forces and moments denoted by $\mathbf{P}=[P_1, \cdots ,P_m]^T$ are represented with reference to the $\mathbf{p}$ coordinates so that $\mathbf{p}$ and $\mathbf{P}$ are the generalized coordinates and their corresponding generalized forces. The problem is to transform the forces and moments denoted by $\mathbf{P}$ from $\mathbf{p}$-coordinates to $\mathbf{q}$-coordinates.

We consider again virtual displacements $\delta\mathbf{p}$. Since the $\mathbf{q}$-coordinates are assumed to be a complete set of generalized coordinates, they can represent the displacement of an arbitrary point in the system. Therefore, displacements represented by $\mathbf{p}$ must be expressed by functions of the $\mathbf{q}$-coordinates. Differentiating the functions, we can relate the virtual displacements $\delta\mathbf{p}$ to $\delta\mathbf{q}$ so that

$$\delta\mathbf{p} = \mathbf{J}\delta\mathbf{q} \tag{4-13}$$

where $\mathbf{J}$ is the $m \times n$ Jacobian matrix associated with the coordinate transformation. We now show that generalized forces $\mathbf{P}$ are transformed to generalized forces $\mathbf{Q}$ in the $\mathbf{q}$-coordinates by

$$\mathbf{Q} = \mathbf{J}^T\mathbf{P} \tag{4-14}$$

where $\mathbf{J}^T$ is the transpose of the Jacobian matrix.

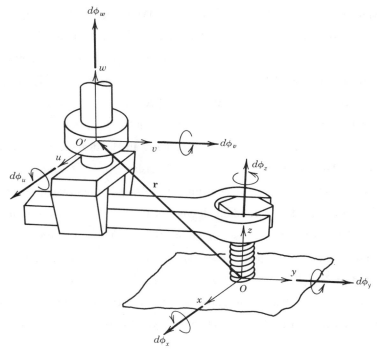

Figure 4-7 :   Example 4-2.

Equation (4-14) can be proved in the same way as before. Namely, we can find the force **Q** equivalent to the force **P** by considering the balance of the two sets of forces **Q** and -**P**. The system is in equilibrium if, and only if, the virtual work done by the external force -**P** and the equivalent force **Q** vanishes for arbitrary virtual displacements that conform to the geometrical relationship (4-13). Since

$$\delta Work = \mathbf{Q}^T \delta \mathbf{q} - \mathbf{P}^T \delta \mathbf{p} = (\mathbf{Q} - \mathbf{J}^T \mathbf{P})^T \delta \mathbf{q} \tag{4-15}$$

The matrix $(\mathbf{Q} - \mathbf{J}^T \mathbf{P})$ must be zero in order for δWork to vanish for an arbitrary $\delta \mathbf{q}$, which thus proves (4-14).

Example 4-2

Figure 4-7 shows the robot hand rotating a screw using a wrench held in its gripper. To perform this task, the force and moment acting on the screw must be monitored. The robot has a force sensor for this purpose, which measures the 6-axis force and moment at the wrist. The problem is to determine the force and moment acting on the screw from the measurement at the wrist. The force and moment are represented in two ways: with reference to the coordinate frame at the screw, $0-xyz$, and with reference to the frame at the wrist force sensor, $0'-uvw$. The two coordinate frames are parallel at the instant shown, and the origin of the wrist coordinate frame $0'$ is given by the 3×1 position vector $\mathbf{r} = [r_x, r_y, r_z]^T$, defined with respect to $0-xyz$.

In this example, we can regard the robot hand, the wrench, and the screw, as parts of a single rigid body. Infinitesimal translations and rotations of the rigid body are represented by six-dimensional vectors $d\mathbf{q} = [dx, dy, dz, d\phi_x, d\phi_y, d\phi_z]^T$ with respect to the screw coordinate frame $0-xyz$ and by another vector $d\mathbf{p} = [du, dv, dw, d\phi_u, d\phi_v, d\phi_w]^T$ with respect to $0'-uvw$. From the figure the coordinate transformation from $d\mathbf{q}$ to $d\mathbf{p}$ is given by

$$
d\mathbf{p} =
\begin{bmatrix} du \\ dv \\ dw \\ d\phi_u \\ d\phi_v \\ d\phi_w \end{bmatrix}
=
\begin{bmatrix}
1 & 0 & 0 & 0 & r_z & -r_y \\
0 & 1 & 0 & -r_z & 0 & r_x \\
0 & 0 & 1 & r_y & -r_x & 0 \\
& & & 1 & 0 & 0 \\
& \mathbf{O} & & 0 & 1 & 0 \\
& & & 0 & 0 & 1
\end{bmatrix}
\begin{bmatrix} dx \\ dy \\ dz \\ d\phi_x \\ d\phi_y \\ d\phi_z \end{bmatrix}
= \mathbf{J}d\mathbf{q}
\qquad (4-16)
$$

In accordance with the infinitesimal displacement vectors, let us write the forces and moments by six-dimensional vector $\mathbf{Q} = [F_x, F_y, F_z, M_x, M_y, M_z]^T$ with respect to $0-xyz$ and by $\mathbf{P} = [F_u, F_v, F_w, M_u, M_v, M_w]^T$ with respect to $0'-uvw$. Applying the theorem, we now find the relationship between $\mathbf{Q}$ and $\mathbf{P}$. From (4-14), we have

$$
\mathbf{Q} =
\begin{bmatrix} F_x \\ F_y \\ F_z \\ M_x \\ M_y \\ M_z \end{bmatrix}
=
\begin{bmatrix}
1 & 0 & 0 & & & \\
0 & 1 & 0 & & \mathbf{O} & \\
0 & 0 & 1 & & & \\
0 & -r_z & r_y & 1 & 0 & 0 \\
r_z & 0 & -r_x & 0 & 1 & 0 \\
-r_y & r_x & 0 & 0 & 0 & 1
\end{bmatrix}
\begin{bmatrix} F_u \\ F_v \\ F_w \\ M_u \\ M_v \\ M_w \end{bmatrix}
\qquad (4-17)
$$

The above equation gives the transformation of the wrist force and moment to the screw force and moment.

△△△

4.2. Stiffness

4.2.1. Introduction

In this section, we analyze the stiffness of a manipulator arm. When a force is applied at the endpoint of a manipulator arm, the endpoint will deflect by an amount which depends on the stiffness of the arm and the force applied. The stiffness of the arm's endpoint determines the strength of the manipulator arm and, more importantly, the positioning accuracy in the presence of disturbance forces and loads. Also, as detailed in Chapter 7, stiffness is an important control variable which allows a robot to perform complex tasks. With the appropriate stiffness, the robot can accommodate endpoint forces with acceptable displacements. In this chapter, we introduce the fundamental concepts and properties of the stiffness of a manipulator arm.

There are several sources that produce deflections of a manipulator arm. Arm links, for example, may deflect when a large force is applied. In particular, as the arm length gets longer, as in the space shuttle manipulator (Nguyen and Ravidran, 1977), the deflection resulting from the link compliance is a major source of the endpoint deflection. In the majority of today's industrial robots, however, the major source of the deflection occurs in transmissions, reducers, and servo drive systems (Sweet and Good, 1985). Each joint is driven by an individual actuator through a reducer and transmission mechanisms. When a drive force or torque is transmitted, each member involved may deflect. Also, the actuator itself has a limited stiffness determined by its feedback control system, which generates the drive torque based on the discrepancy between the reference position and the actual measured position. The stiffness of the drive system is then dependent on the loop gain of the feedback system. We model the stiffness of the drive system combined with the stiffness of the reducer and transmissions by a spring constant k_i that relates the deflection at joint i to the force or torque transmitted. Namely,

$$\tau_i = k_i \Delta q_i \qquad\qquad\qquad (4-18)$$

where τ_i is the joint torque and Δq_i is the deflection at the joint axis. In the following analysis, we assume that the arm links are rigid, and investigate the endpoint stiffness based upon the model of the joint stiffness given by (4-18).

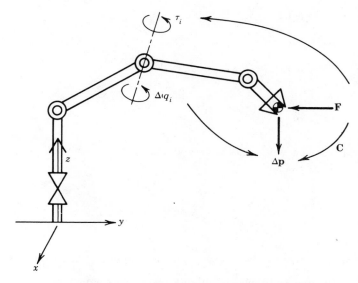

Figure 4-8 :   Endpoint compliance and joint servo stiffness.

4.2.2. Endpoint Compliance Analysis

In this section, we derive the endpoint stiffness from the individual joint stiffnesses. As shown in Figure 4-8, we denote the endpoint force and moment by the m-dimensional vector **F** and the resultant deflection by $\Delta \mathbf{p}$, both of them defined with reference to the base coordinate frame. When we neglect gravity and friction at the joints, the endpoint force can be converted to the equivalent joint torques according to the theorem in Section 4.1. Namely,

$$\boldsymbol{\tau} = \mathbf{J}^T \mathbf{F} \tag{4-19}$$

where $\mathbf{J}^T$ is the $n \times m$ transpose of the manipulator Jacobian. At the individual joints, joint torques $\boldsymbol{\tau}$ are related to joint deflections $\Delta \mathbf{q}$ by the individual stiffnesses as we modeled in the previous section. For convenience, let us rewrite (4-18) in vector form:

$$\boldsymbol{\tau} = \mathbf{K} \Delta \mathbf{q} \tag{4-20}$$

where **K** is a $n \times n$ diagonal matrix given by

$$K = \begin{bmatrix} k_1 & & O \\ & \cdot & \\ & & \cdot \\ O & & k_n \end{bmatrix}$$ (4–21)

From Section 3.1, the individual joint deflections Δq produce the endpoint deflection Δp according to

$$\Delta p = J \Delta q$$ (4–22)

When the individual joint drive systems are active and the stiffnesses are non-zero, the matrix **K** is invertible. Substituting (4-19) and (4-20) into (4-22), we obtain

$$\Delta p = CF$$ (4–23)

where

$$C = JK^{-1}J^T$$ (4–24)

Thus the deflection at the endpoint Δp is related to the endpoint force **F** by the $m \times m$ matrix **C**. The matrix **C** is called the *compliance matrix* of the arm endpoint.

If the manipulator Jacobian is a square matrix and of full rank, the compliance matrix is invertible:

$$F = C^{-1}\Delta p$$ (4–25)

The inverse of the compliance matrix is called the *stiffness matrix* of the arm endpoint. When the manipulator Jacobian is degenerate, the stiffness becomes infinite in at least one direction. According to the linear mapping diagrams in Figure 4-6, space S_2 is not reduced to 0 when the Jacobian matrix is degenerate. This implies that there exists a null space $N(J^T)$ in which the endpoint force is mapped into zero joint torques. Therefore, if the endpoint force acts in the direction involved in the null space $N(J^T)$, no joint torques are induced, hence no joint

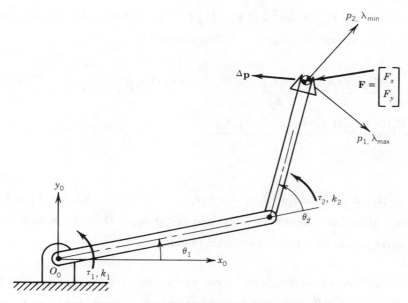

Figure 4-9 :   Principal directions of endpoint compliance.

deflections. As a result, no endpoint deflection occurs, so that stiffness is infinite (if the arm links are assumed to be rigid).

The endpoint compliance matrix as well as the stiffness matrix consist of the individual joint stiffnesses and the manipulator Jacobian. Since the Jacobian varies with the arm configuration, the compliance matrix is configuration-dependent. Also, at a given arm configuration, the magnitude of the endpoint deflection varies with the direction of the endpoint force.

4.2.3. The Principal Transformation of Compliance Matrices

As mentioned in the previous section, the endpoint deflection of a manipulator arm varies depending not only on the arm configuration but also on the direction of the endpoint force applied. In this section, we analyze the maximum and minimum deflections of the arm's endpoint and characterize the compliance matrix.

To simplify the analysis, we deal with the two degree-of-freedom planar manipulator shown in Figure 4-9. The endpoint deflection and endpoint force are represented by the two-dimensional vectors $\Delta\mathbf{p} = [\Delta x, \Delta y]^T$ and $\mathbf{F} = [F_x, F_y]^T$, respectively. We begin by deriving the endpoint compliance matrix from equation (4-24), namely

$$
\mathbf{C} = \begin{bmatrix}
\dfrac{(l_1 s_1 + l_2 s_{12})^2}{k_1} + \dfrac{l_2^2 s_{12}^2}{k_2} & \dfrac{-(l_1 c_1 + l_2 c_{12})(l_1 s_1 + l_2 s_{12})}{k_1} - \dfrac{l_2^2 c_{12} s_{12}}{k_2} \\[3mm]
\dfrac{-(l_1 c_1 + l_2 c_{12})(l_1 s_1 + l_2 s_{12})}{k_1} - \dfrac{l_2^2 c_{12} s_{12}}{k_2} & \dfrac{(l_1 c_1 + l_2 c_{12})^2}{k_1} + \dfrac{l_2^2 c_{12}^2}{k_2}
\end{bmatrix} \quad (4-26)
$$

where $c_1 = \cos(\theta_1)$, $c_{12} = \cos(\theta_1 + \theta_2)$, $s_1 = \sin(\theta_1)$, $s_{12} = \sin(\theta_1 + \theta_2)$, and k_1 and k_2 are the individual joint stiffnesses. Equations (4-21) and (4-24) imply that the compliance matrix is always symmetric, as can be verified in equation (4-26).

For the compliance matrix obtained above and a given arm configuration, let us find the maximum and minimum deflections and their directions when a unit magnitude force is applied to the endpoint. From (4-23), the squared norm of the endpoint deflection is given by

$$
|\Delta\mathbf{p}|^2 = \Delta\mathbf{p}^T \Delta\mathbf{p} = \mathbf{F}^T \mathbf{C}^T \mathbf{C} \mathbf{F} = \mathbf{F}^T \mathbf{C}^2 \mathbf{F} \tag{4-27}
$$

where $\mathbf{C}$ is symmetric. We evaluate the maximum and minimum under the condition on the magnitude of the endpoint force:

$$
|\mathbf{F}|^2 = \mathbf{F}^T \mathbf{F} = 1 \tag{4-28}
$$

To solve this problem, we employ Lagrange multiplier λ to define

$$
L = \mathbf{F}^T \mathbf{C}^2 \mathbf{F} - \lambda(\mathbf{F}^T \mathbf{F} - 1) \tag{4-29}
$$

The necessary condition for the squared norm of the endpoint deflection to take extreme values is given by

$$
\frac{\partial L}{\partial \lambda} = 0 : \quad -\mathbf{F}^T \mathbf{F} + 1 = 0 \tag{4-30}
$$

which is identical to (4-28), and

$$\frac{\partial L}{\partial \mathbf{F}} = \mathbf{0}: \quad \mathbf{C}^2\mathbf{F} - \lambda\mathbf{F} = \mathbf{0} \tag{4-31}$$

From equation (4-31), it follows that the Lagrange multiplier is the eigenvalue of the squared compliance matrix $\mathbf{C}^2$. Thus the problem of finding the maximum and minimum deflections is basically an eigenvalue problem. Solving the characteristic equation for $\mathbf{C}^2$ yields the maximum and minimum eigenvalues

$$\begin{matrix} \lambda_{max} \\ \lambda_{min} \end{matrix} = \frac{1}{2}\left[a_1 + a_2 \pm \sqrt{(a_1-a_2)^2 + 4a_3^2} \right] \tag{4-32}$$

where

$$\mathbf{C}^2 = \begin{bmatrix} a_1 & a_3 \\ a_3 & a_2 \end{bmatrix}$$

Note that both eigenvalues are positive, since the individual joint stiffnesses are positive. Using the eigenvalues and equations (4-30) and (4-31), the squared norm of the endpoint deflection is given by

$$|\Delta\mathbf{p}|^2 = \mathbf{F}^T\mathbf{C}^2\mathbf{F} = \mathbf{F}^T \lambda\mathbf{F} = \lambda \tag{4-33}$$

Thus, the maximum and minimum deflections are given by $\sqrt{\lambda_{max}}$ and $\sqrt{\lambda_{min}}$, respectively.

The direction in which the maximum or minimum deflection occurs is given by the eigenvector corresponding to the maximum or minimum eigenvalue. Figure 4-9 illustrates the directions of the eigenvectors. Note that the two directions are orthogonal to each other. These directions are referred to as *principal directions*. Let us define coordinate axes in the principal directions and call them *principal axes*. The compliance matrix becomes diagonal when expressed in the principal coordinates. Let $\mathbf{e}_1$ and $\mathbf{e}_2$ be unit vectors along the principal axes, associated repectively with the maximum and minimum eigenvalues; and let $\mathbf{E}$ be a 2 $\times$ 2 matrix consisting of $\mathbf{e}_1$ and $\mathbf{e}_2$:

$$\mathbf{E} = [\mathbf{e}_1 \quad \mathbf{e}_2]$$

The compliance matrix is then transformed to the diagonalized form $\mathbf{C}^*$ in the principal coordinates:

$$\mathbf{C}^* = \mathbf{E}^T \mathbf{C} \mathbf{E} = \left[\begin{array}{cc} \sqrt{\lambda_{max}} & 0 \\ 0 & \sqrt{\lambda_{min}} \end{array} \right] \tag{4-34}$$

where $\mathbf{E}^T = \mathbf{E}^{-1}$ since $\mathbf{E}$ is orthonormal.

The coordinate transformation to the principal coordinates is referred to as the *principal transformation*. When the endpoint force is applied in the principal direction, the deflection occurs also in the same principal direction and the magnitude of the deflection takes an extreme value.

4.3. Research Topics

The static force analysis and control problems have been extended to mechanical systems containing closed-loop kinematic chains. Such systems include manipulators with parallel drive mechanisms (Asada and Youcef-Toumi, 1983, 1984), the coordinated motion control of two arms carrying the same object together (Nakano, *et.al.*, 1974), multiple legs contacting the ground (Orin and Oh, 1981), multiple fingers grasping an object (Salisbury, 1982), and manipulators braced against the environment surface (West and Asada, 1985). In these systems, the motion of each mechanism is constrained by the kinematic loops involved, so that internal forces are present inside the kinematic loops. It is necessary to control these internal forces, in addition to controlling the external forces exerted on the mechanism.

Compliance is an important characteristic when manipulators mechanically interact with their environment, as will be further discussed in Chapter 7. In particular, precision part mating in assembly operations can be made possible for crude robot manipulators by appropriately incorporating compliance at their end-effectors (Drake, 1977). The "Remote-Center-Compliance" hand, a device specially designed to this effect, has been extensively studied and adapted for practical use (Whitney, 1982; Nevins and Whitney, *et.al.*, 1974-1977; Arai and Kinoshita, 1981). (Hanafusa and Asada, 1977) addressed grasping compliance and stability using potential functions.

Chapter 5
DYNAMICS

In this chapter, we analyze the dynamic behavior of manipulator arms. The dynamic behavior is described in terms of the time rate of change of the arm configuration in relation to the joint torques exerted by the actuators. This relationship can be expressed by a set of differential equations, called *equations of motion*, that govern the dynamic response of the arm linkage to input joint torques. In the next chapter, we will design a control system on the basis of these equations of motion.

Two methods can be used in order to obtain the equations of motion: the *Newton-Euler formulation*, and the *Lagrangian formulation*. The Newton-Euler formulation is derived by the direct interpretation of Newton's Second Law of Motion, which describes dynamic systems in terms of force and momentum. The equations incorporate all the forces and moments acting on the individual arm links, including the coupling forces and moments between the links. The equations obtained from the Newton-Euler method include the constraint forces acting between adjacent links. Thus, additional arithmetic operations are required to eliminate these terms and obtain explicit relations between the joint torques and the resultant motion in terms of joint displacements. In the Lagrangian formulation, on the other hand, the system's dynamic behavior is described in terms of work and energy using generalized coordinates. All the workless forces and constraint forces are automatically eliminated in this method. The resultant equations are generally compact and provide a closed-form expression in terms of joint torques and joint displacements. Further, the derivation is simpler and more systematic than in the Newton-Euler method.

The manipulator's equations of motion are basically a description of the relationship between the input joint torques and the output motion, i.e. the motion of the arm linkage. As in kinematics and in statics, we need to solve the inverse problem of finding the necessary input torques to obtain a desired output motion. This *inverse dynamics* problem is discussed in the last section of this chapter. Recently, efficient algorithms have been developed that allow the dynamic computations to be carried out on-line in real time.

5.1. Newton-Euler Formulation of Equations of Motion

5.1.1. Basic Dynamic Equations

In this section we derive the equations of motion for an individual arm link. As discussed in Chapters 2 and 3, the motion of a rigid body can be decomposed into the translational motion of an arbitrary point fixed to the rigid body, and the rotational motion of the rigid body about that point. The dynamic equations of a rigid body can also be represented by two equations: one describes the translational motion of the centroid (or center of mass), while the other describes the rotational motion about the centroid. The former is Newton's equation of motion for a mass particle, and the latter is called Euler's equation of motion.

We begin by considering the free body diagram of an individual arm link. Figure 5-1 shows all the forces and moments acting on any given link i. The figure is the same as Figure 4-1, which describes the static balance of forces, except for the inertial force and moment that arise from the dynamic motion of the link. Let $\mathbf{v}_{ci}$ be the linear velocity of the centroid of link i with reference to the base coordinate frame $O_0 - x_0 y_0 z_0$, which is an inertial reference frame. The inertial force is then given by $-m_i \dot{\mathbf{v}}_{ci}$, where m_i is the mass of the link and $\dot{\mathbf{v}}_{ci}$ is the time derivative of $\mathbf{v}_{ci}$. The equation of motion is then obtained by adding the inertial force to the static balance of forces in equation (4-1) so that

$$\mathbf{f}_{i-1,i} - \mathbf{f}_{i,i+1} + m_i \mathbf{g} - m_i \dot{\mathbf{v}}_{ci} = 0 \qquad i = 1, \cdots, n \tag{5-1}$$

where, as in Chapter 4, $\mathbf{f}_{i-1,i}$ and $-\mathbf{f}_{i,i+1}$ are the coupling forces applied to link i by links $i-1$ and $i+1$, respectively, and $\mathbf{g}$ is the acceleration of gravity.

Rotational motions are described by Euler's equations. In the same way as for translational motions, the dynamic equations are derived by adding "inertial torques" to the

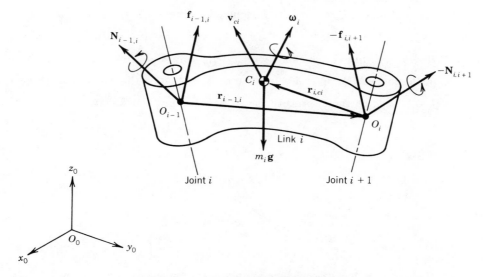

Figure 5-1 :   Free body diagram of link i.

static balance of moments. We begin by describing the mass properties of a single rigid body with respect to rotations about the centroid. The mass properties are represented by an *inertia tensor*, which is a 3 $\times$ 3 symmetric matrix defined by

$$
\mathbf{I} = \begin{bmatrix} \int \{(y-y_c)^2+(z-z_c)^2\}\rho dV & -\int (x-x_c)(y-y_c)\rho dV & -\int (z-z_c)(x-x_c)\rho dV \\ -\int (x-x_c)(y-y_c)\rho dV & \int \{(z-z_c)^2+(x-x_c)^2\}\rho dV & -\int (y-y_c)(z-z_c)\rho dV \\ -\int (z-z_c)(x-x_c)\rho dV & -\int (y-y_c)(z-z_c)\rho dV & \int \{(x-x_c)^2+(y-y_c)^2\}\rho dV \end{bmatrix}
$$

$$(5-2)$$

where ρ is the mass density, x_c, y_c, and z_c are the coordinates of the centroid of the rigid body, and each integral is taken over the entire volume V of the rigid body. Note that the inertia tensor varies with the orientation of the rigid body.

The inertial torque acting on link i is given by the time rate of change of the angular momentum of the link at that instant. Let $\boldsymbol{\omega}_i$ be the angular velocity vector and $\mathbf{I}_i$ be the centroidal inertia tensor of link i ; then the angular momentum is given by $\mathbf{I}_i\,\boldsymbol{\omega}_i$. Since the inertia tensor varies as the orientation of the link changes, the time derivative of the angular momentum includes not only the angular acceleration term $\mathbf{I}_i\,\dot{\boldsymbol{\omega}}_i$, but also a term resulting from changes in the inertia tensor. This latter term is known as the *gyroscopic torque* and is given by $\boldsymbol{\omega}_i\times(\mathbf{I}_i\,\boldsymbol{\omega}_i)$. Adding these terms to the original balance of moments (4-2) yields

$$\mathbf{N}_{i-1,i} - \mathbf{N}_{i,i+1} + \mathbf{r}_{i,ci}\times\mathbf{f}_{i,i+1} - \mathbf{r}_{i-1,ci}\times\mathbf{f}_{i-1,i} - \mathbf{I}_i\,\dot{\boldsymbol{\omega}}_i - \boldsymbol{\omega}_i\times(\mathbf{I}_i\,\boldsymbol{\omega}_i) \,=\, 0 \qquad (5\text{--}3)$$
$$i = 1, \cdots, n$$

using the notations of Figure 4-1.

Equations (5-1) and (5-3) govern the dynamic behavior of an individual arm link. The complete set of equations for the whole manipulator arm is obtained by evaluating both equations for all the arm links, $i = 1, \cdots, n$.

5.1.2. Closed-Form Dynamic Equations

The Newton-Euler equations we have derived are not in an appropriate form for use in dynamic analysis and controller design. They do not explicitly describe the input-output relationship, unlike the relationships we obtained for kinematics and statics. In this section, we modify the Newton-Euler equations so that explicit input-output relations can be obtained.

The Newton-Euler equations involve coupling forces and moments $\mathbf{f}_{i-1,i}$ and $\mathbf{N}_{i-1,i}$. As shown in equations (4-4) or (4-5), the joint torque τ_i , which is the input to the arm linkage, is included in the coupling force or moment. However, τ_i is not *explicitly* involved in the Newton-Euler equations. The coupling force and moment also include workless constraint forces, which act *internally* so that individual link motions conform to the geometric constraints imposed by the arm linkage. To derive explicit input-output dynamic relations, we need to separate the input joint torques from the constraint forces and moments.

The Newton-Euler equations are described in terms of centroid velocities and accelerations of individual arm links. Individual link motions, however, are not independent, but are coupled through the arm linkage. They must satisfy certain kinematic relationships to

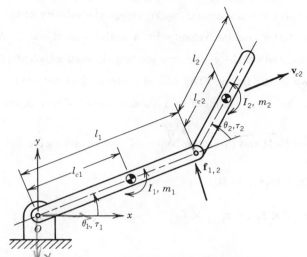

Figure 5-2 :   Mass properties of two d.o.f. planar manipulator.

conform to the geometric constraints. Thus, individual centroid position variables are not appropriate for output variables since they are not independent.

The appropriate form of the dynamic equations therefore consists of equations described in terms of all independent position variables and input forces, i.e., joint torques, that are explicitly involved in the dynamic equations. Dynamic equations in such an explicit input-output form are referred to as *closed-form dynamic equations*. As discussed in the previous chapter, joint displacements **q** are a complete and independent set of generalized coordinates that locate the whole arm linkage, and joint torques τ are a set of independent inputs that are separated from constraint forces and moments. Hence, dynamic equations in terms of joint displacements **q** and joint torques τ are closed-form dynamic equations.

Example 5-1

Figure 5-2 shows the two degree-of-freedom planar manipulator that we discussed in the previous chapter. Let us obtain the Newton-Euler equations of motion for the two individual links, and then derive the closed-form dynamic equations in terms of joint displacements θ_1 and θ_2 , and joint torques τ_1 and τ_2 .

Since the link mechanism is planar, we represent the velocity of the centroid of each link by a 2-vector $\mathbf{v}_{ci}$ and the angular velocity by a scalar velocity ω_i. We assume that the centroid of link i is located on the center line passing through adjacent joints at a distance l_{ci} from joint i, as shown in the figure. The axis of rotation does not vary for the planar linkage. The inertia tensor in this case is reduced to a scalar moment of inertia denoted by I_i.

From equations (5-1) and (5-3), the Newton-Euler equations for link 1 are given by

$$\mathbf{f}_{0,1} - \mathbf{f}_{1,2} + m_1\mathbf{g} - m_1\dot{\mathbf{v}}_{c1} = \mathbf{0}$$

$$N_{0,1} - N_{1,2} + \mathbf{r}_{1,c1} \times \mathbf{f}_{1,2} - \mathbf{r}_{0,c1} \times \mathbf{f}_{0,1} - I_1\dot{\omega}_1 = 0 \tag{5-4}$$

Note that all vectors are 2×1, so that the $N_{i-1,i}$ and the vector products are scalar quantities. Similarly, for link 2,

$$\mathbf{f}_{1,2} + m_2\mathbf{g} - m_2\dot{\mathbf{v}}_{c2} = \mathbf{0}$$

$$N_{1,2} + \mathbf{r}_{1,c2} \times \mathbf{f}_{1,2} - I_2\dot{\omega}_2 = 0 \tag{5-5}$$

To obtain closed-form dynamic equations, we first eliminate the constraint forces and separate them from the joint torques, so as to explicitly involve the joint torques in the dynamic equations. For the planar manipulator, the joint torques τ_1 and τ_2 are equal to the coupling moments:

$$N_{i-1,i} = \tau_i \tag{5-6}$$

Substituting (5-6) into (5-5) and eliminating $\mathbf{f}_{1,2}$, we obtain

$$\tau_2 - \mathbf{r}_{1,c2} \times m_2\dot{\mathbf{v}}_{c2} + \mathbf{r}_{1,c2} \times m_2\mathbf{g} - I_2\dot{\omega}_2 = 0 \tag{5-7}$$

Similarly, eliminating $\mathbf{f}_{0,1}$ yields

$$\tau_1 - \tau_2 - \mathbf{r}_{0,c1} \times m_1\dot{\mathbf{v}}_{c1} - \mathbf{r}_{0,1} \times m_2\dot{\mathbf{v}}_{c2} + \mathbf{r}_{0,c1} \times m_1\mathbf{g} + \mathbf{r}_{0,1} \times m_2\mathbf{g} - I_1\dot{\omega}_1 = 0 \tag{5-8}$$

Next, we rewrite $\mathbf{v}_{ci}$, ω_i, and $\mathbf{r}_{i,i+1}$ using joint displacements θ_1 and θ_2, which are independent variables. Note that ω_2 is the angular velocity relative to the base coordinate frame, while $\dot{\theta}_2$ is measured relative to link 1. Then, we have

$$\omega_1 = \dot{\theta}_1 \qquad \omega_2 = \dot{\theta}_1 + \dot{\theta}_2 \tag{5--9}$$

The linear velocities can be written as

$$\mathbf{v}_{c1} = \left[\begin{array}{c} -l_{c1}\,\dot{\theta}_1 \sin(\theta_1) \\[2ex] l_{c1}\,\dot{\theta}_1 \cos(\theta_1) \end{array} \right] \tag{5-10}$$

$$\mathbf{v}_{c2} = \left[\begin{array}{c} -\{l_1 \sin(\theta_1) + l_{c2}\sin(\theta_1+\theta_2)\}\dot{\theta}_1 - l_{c2}\sin(\theta_1+\theta_2)\,\dot{\theta}_2 \\[2ex] \{l_1 \cos\theta_1 + l_{c2}\cos(\theta_1+\theta_2)\}\dot{\theta}_1 + l_{c2}\cos(\theta_1+\theta_2)\,\dot{\theta}_2 \end{array} \right]$$

Substituting equations (5-9) and (5-10) along with their time derivatives into equations (5-7) and (5-8), we obtain the closed-form dynamic equations in terms of θ_1 and θ_2 :

$$\tau_1 = H_{11}\ddot{\theta}_1 + H_{12}\ddot{\theta}_2 - h\dot{\theta}_2^2 - 2\,h\dot{\theta}_1\dot{\theta}_2 + G_1 \tag{5--11--a}$$

$$\tau_2 = H_{22}\ddot{\theta}_2 + H_{12}\ddot{\theta}_1 + h\dot{\theta}_1^2 + G_2 \tag{5--11--b}$$

where

$$H_{11} = m_1\,l_{c1}^2 + I_1 + m_2[l_1^2 + l_{c2}^2 + 2\,l_1 l_{c2}\cos(\theta_2)] + I_2 \tag{5--12--a}$$

$$H_{22} = m_2\,l_{c2}^2 + I_2 \tag{5--12--b}$$

$$H_{12} = m_2\,l_1 l_{c2}\cos(\theta_2) + m_2 l_{c2}^2 + I_2 \tag{5--12--c}$$

$$h = m_2\,l_1 l_{c2}\sin(\theta_2) \tag{5--12--d}$$

$$G_1 = m_1\,l_{c1}\,g\cos(\theta_1) + m_2 g\{l_{c2}\cos(\theta_1+\theta_2) + l_1\cos(\theta_1)\} \tag{5--12--e}$$

$$G_2 = m_2\,l_{c2}\,g\cos(\theta_1+\theta_2) \tag{5--12--f}$$

The scalar g represents the acceleration of gravity along the negative y axis.   △△△

More generally, the closed-form dynamic equations of an n-degree-of-freedom manipulator can be given in the form

$$\tau_i = \sum_{j=1}^{n} H_{ij} \, \ddot{q}_j + \sum_{j=1}^{n} \sum_{k=1}^{n} h_{ijk} \, \dot{q}_j \dot{q}_k + G_i \qquad\qquad i = 1, \cdots, n \qquad\qquad (5-13)$$

where coefficients H_{ij}, h_{ijk} and G_i are functions of joint displacements $q_1, \cdots, q_n$. When external forces act on the manipulator arm, the left-hand side must be modified accordingly.

5.1.3. Physical Interpretation of the Dynamic Equations

In this section, we interpret the physical meaning of each term involved in the closed-form dynamic equations for the two degree-of-freedom planar manipulator.

The last term G_i in each of equations (5-11-a,b) accounts for the effect of gravity. Indeed, the terms G_1 and G_2, given by (5-12-e,f), represent the moments created by the masses m_1 and m_2 about their individual joint axes. The moments are dependent upon the arm configuration. When the arm is fully extended along the x axis, the gravity moments are maximum.

Next, we investigate the first terms in the dynamic equations. When the second joint is immobilized, i.e. $\ddot{\theta}_2 = 0$ and $\dot{\theta}_2 = 0$, the first dynamic equation reduces to $\tau_1 = H_{11} \ddot{\theta}_1$, where the gravity term is neglected. From this expression it follows that the coefficient H_{11} accounts for the moment of inertia seen by the first joint when the second joint is immobilized. The coefficient H_{11} given by equation (5-12-a) is interpreted as the total moment of inertia of both links reflected to the first joint axis. The first two terms, $m_1 l_{c1}^2 + I_1$, in equation (5-12-a), represent the moment of inertia of link 1 with respect to joint 1, while the other terms are the contribution from link 2. The inertia of the second link depends upon the distance L between the centroid of link 2 and the first joint axis shown in Figure 5-3. The distance L is a function of the joint angle θ_2 and is given by

$$L^2 = l_1^2 + l_{c2}^2 + 2 \, l_1 l_{c2} \cos(\theta_2) \qquad\qquad (5-14)$$

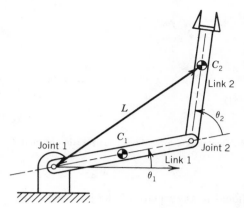

Figure 5-3 :   Varying inertia depending on arm configuration.

Using the parallel axes theorem of inertia tensors (Goldstein, 1981), the inertia of link 2 with respect to joint 1 is $m_2L^2+I_2$, which is consistent with the last two terms in equation (5-12-a). Note that the inertia varies with the arm configuration. The inertia is maximum when the arm is fully extended ($\theta_2 = 0$), and minimum when the arm is completely contracted ($\theta_2 = \pi$).

Let us now investigate the second terms in equation (5-11). Consider the instant when $\dot{\theta}_1 = \dot{\theta}_2 = 0$ and $\ddot{\theta}_1 = 0$, then the first equation reduces to $\tau_1 = H_{12}\ddot{\theta}_2$, where the gravity term is again neglected. From this expression it follows that the second term accounts for the effect of the second link motion upon the first joint. When the second link is accelerated, the reaction force and torque induced by the second link act upon the first link. This is clear in the original Newton-Euler equations (5-4), where the coupling force $-\mathbf{f}_{1,2}$ and moment $-\mathbf{N}_{1,2}$ from link 2 are involved in the dynamic equation for link 1. The coupling force and moment cause a torque τ_{int} about the first joint axis given by

$$
\begin{aligned}
\tau_{int} &= -N_{1,2} - \mathbf{r}_{0,1} \times \mathbf{f}_{1,2} \\
&= -I_2\dot{\omega}_2 - \mathbf{r}_{0,c2} \times m_2\dot{\mathbf{v}}_{c2} \\
&= -[I_2 + m_2(l_{c2}^2 + l_1l_{c2}\cos\theta_2)]\ddot{\theta}_2
\end{aligned}
\tag{5-15}
$$

where $N_{1,2}$ and $\mathbf{f}_{1,2}$ are evaluated using equation (5-5) for $\dot{\theta}_1 = \dot{\theta}_2 = 0$ and $\ddot{\theta}_1 = 0$. This agrees with the second term in equation (5-11-a). Thus, the second term accounts for the interaction between the two joints.

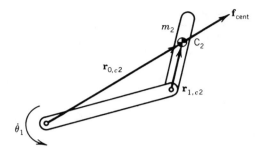

Figure 5-4 : Centrifugal force due to the rotation of joint 1.

The third terms in equation (5-11) are proportional to the square of the joint velocities. We consider the instant when $\dot{\theta}_2 = 0$, and $\ddot{\theta}_1 = \ddot{\theta}_2 = 0$, as shown in Figure 5-4. In this case, a centrifugal force acts upon the second link. Let $\mathbf{f}_{cent}$ be the centrifugal force. Its magnitude is given by

$$|\mathbf{f}_{cent}| = m_2 L \, \dot{\theta}_1^2 \qquad\qquad (5-16)$$

where L is the distance between the centroid C_2 and the first joint O_0. The direction of the centrifugal force is parallel to position vector $\overrightarrow{O_0 C_2}$. This centrifugal force causes a moment τ_{cent} about the second joint. Using equation (5-16), the moment τ_{cent} is computed as

$$\tau_{cent} = \mathbf{r}_{1,c2} \times \mathbf{f}_{cent} = -m_2 \, l_1 l_{c2} \, \dot{\theta}_1^2 \sin(\theta_2) \qquad\qquad (5-17)$$

This agrees with the third term $h\dot{\theta}_1^2$ in equation (5-11-b). Thus we conclude that the third term is caused by the centrifugal effect on the second joint due to the motion of the first joint. Similarly, when the second joint is rotated at a constant velocity $\dot{\theta}_2$, the torque caused by the centrifugal effect acts upon the first joint.

Finally we discuss the fourth term of equation (5-11-a), which is proportional to the product of the joint velocities. Consider the instant when the two joints rotate at velocities $\dot{\theta}_1$ and $\dot{\theta}_2$ at the same time. Let $O_b-x_b y_b$ be the coordinate frame attached to the tip of link 1, as shown in Figure 5-5. Note that the frame $O_b-x_b y_b$ is parallel to the base coordinate frame at

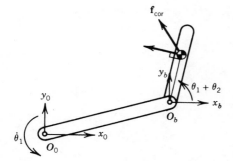

Figure 5-5 : Coriolis effect.

the instant shown. However, the frame rotates at the angular velocity $\dot{\theta}_1$ along with link 1. The motion of link 2 is represented by $\dot{\theta}_2$, relative to link 1 or the moving coordinate frame $O_b-x_by_b$. When a mass particle m moves at a velocity of $\mathbf{v}_b$ relative to a moving coordinate frame rotating at an angular velocity $\boldsymbol{\omega}$, the mass particle has the so-called *Coriolis force* given by $2m(\boldsymbol{\omega}\times\mathbf{v}_b)$. Let $\mathbf{f}_{Cor}$ be the force acting on link 2 due to the Coriolis effect. The Coriolis force is given by

$$\mathbf{f}_{Cor}=\begin{bmatrix} -2\,m_2\,l_{c2}\,\dot{\theta}_1\dot{\theta}_2\,\cos\,(\theta_1+\theta_2) \\ -2\,m_2\,l_{c2}\,\dot{\theta}_1\dot{\theta}_2\,\sin\,(\theta_1+\theta_2) \end{bmatrix} \tag{5-18}$$

This Coriolis force causes a moment τ_{Cor} about the first joint, which is given by

$$\tau_{Cor}=\mathbf{r}_{1,c2}\times\mathbf{f}_{Cor}=2\,m_2\,l_1l_{c2}\,\dot{\theta}_1\dot{\theta}_2\,\sin\,(\theta_2) \tag{5-19}$$

The right-hand side of the above equation agrees with the fourth term in equation (5-11-a). Since the Coriolis force given by equation (5-18) acts in parallel with link 2, the force does not create a moment about the second joint in this particular case.

Thus, the dynamic equations of a manipulator arm are characterized by a configuration-dependent inertia, gravity torques, and interaction torques caused by the accelerations of the other joints and the existence of centrifugal and Coriolis effects.

5.2. Lagrangian Formulation of Manipulator Dynamics

5.2.1. Lagrangian Dynamics

In the Newton-Euler formulation, the equations of motion are derived from Newton's Second Law, which relates force and momentum, as well as torque and angular momentum. The resulting equations involve constraint forces, which must be eliminated in order to obtain closed-form dynamic equations. In the Newton-Euler formulation, the equations are not expressed in terms of independent variables, and do not include input joint torques explicitly. Arithmetic operations are needed to derive the closed-form dynamic equations. This represents a complex procedure which requires physical intuition, as discussed in the previous section.

An alternative to the Newton-Euler formulation of manipulator dynamics is the Lagrangian formulation, which describes the behavior of a dynamic system in terms of work and energy stored in the system rather than of forces and momenta of the individual members involved. The constraint forces involved in the system are automatically eliminated in the formulation of Lagrangian dynamic equations. The closed-form dynamic equations can be derived systematically in any coordinate system.

Let $q_1, \cdots, q_n$ be generalized coordinates that completely locate a dynamic system. Let T and U be the total kinetic energy and potential energy stored in the dynamic system. We define the Lagrangian $\mathcal{L}$ by

$$\mathcal{L}(q_i, \dot{q}_i) = T - U \tag{5-20}$$

Note that, since the kinetic and potential energies are functions of q_i and $\dot{q}_i$, $(i = 1, \cdots, n)$, so is the Lagrangian $\mathcal{L}$. Using the Lagrangian, equations of motion of the dynamic system are given by

$$\frac{d}{dt} \frac{\partial \mathcal{L}}{\partial \dot{q}_i} - \frac{\partial \mathcal{L}}{\partial q_i} = Q_i \qquad i = 1, \cdots, n \tag{5-21}$$

where Q_i is the generalized force corresponding to the generalized coordinate q_i. The

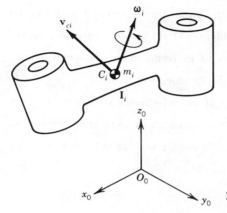

Figure 5-6 :   Centroidal velocity and angular veloity of link i.

generalized force can be identified by considering the virtual work done by non-conservative forces acting on the system.

5.2.2. The Manipulator Inertia Tensor

In this section and the following section, we derive the equations of motion of a manipulator arm using the Lagrangian. We begin by deriving the kinetic energy stored in an individual arm link. As shown in Figure 5-6, let $\mathbf{v}_{ci}$ and ω_i be the 3×1 velocity vector of the centroid and the 3×1 angular velocity vector with reference to the base coordinate frame, which is an inertial reference frame. The kinetic energy of link i is then given by

$$T_i = \frac{1}{2} m_i \mathbf{v}_{ci}^T \mathbf{v}_{ci} + \frac{1}{2} \omega_i^T \mathbf{I}_i \omega_i \qquad (5-22)$$

where m_i is the mass of the link and $\mathbf{I}_i$ is the 3×3 inertia tensor at the centroid expressed in the base coordinates. The first term in the above equation accounts for the kinetic energy resulting from the translational motion of the mass m_i, while the second term represents the kinetic energy resulting from rotation about the centroid. The total kinetic energy stored in the whole arm linkage is then given by

$$T = \sum_{i=1}^{n} T_i \qquad (5-23)$$

since energy is additive.

The expression for the kinetic energy is written in terms of the velocity and angular velocity of each link member, which are not independent variables, as mentioned in the previous section. Let us now rewrite the above equations in terms of an independent and complete set of generalized coordinates, namely joint displacements $q = [q_1, \cdots ,q_n]^T$. In Chapter 3, we analyzed the velocity and angular velocity of an end-effector in relation to joint velocities. We can employ the same method to compute the velocity and angular velocity of an individual link, if we regard the link as an end-effector. Namely, replacing subscripts n and e by i and ci, respectively, in equations (3-19) and (3-23), we obtain

$$\mathbf{v}_{ci} = \mathbf{J}_{L1}^{(i)}\dot{q}_1 + \cdots + \mathbf{J}_{Li}^{(i)}\dot{q}_i = \mathbf{J}_L^{(i)}\dot{\mathbf{q}}$$

$$\boldsymbol{\omega}_i = \mathbf{J}_{A1}^{(i)}\dot{q}_1 + \cdots + \mathbf{J}_{Ai}^{(i)}\dot{q}_i = \mathbf{J}_A^{(i)}\dot{\mathbf{q}}$$

$$(5-24)$$

where $\mathbf{J}_{Lj}^{(i)}$ and $\mathbf{J}_{Aj}^{(i)}$ are the j-th column vectors of the $3 \times n$ Jacobian matrices $\mathbf{J}_L^{(i)}$ and $\mathbf{J}_A^{(i)}$, for linear and angular velocities of link i, respectively. Namely,

$$\mathbf{J}_L^{(i)} = [\mathbf{J}_{L1}^{(i)} \cdots \mathbf{J}_{Li}^{(i)} \; \mathbf{0} \cdots \mathbf{0}]$$

$$\mathbf{J}_A^{(i)} = [\mathbf{J}_{A1}^{(i)} \cdots \mathbf{J}_{Ai}^{(i)} \; \mathbf{0} \cdots \mathbf{0}]$$

$$(5-25)$$

Note that, since the motion of link i depends on only joints 1 through i, the column vectors are set to zero for $j \geq i$. From equations (3-26) and (3-27) each column vector is given by

$$\mathbf{J}_{Lj}^{(i)} = \begin{cases} \mathbf{b}_{j-1} & \text{for a prismatic joint} \\ \mathbf{b}_{j-1} \times \mathbf{r}_{0,ci} & \text{for a revolute joint} \end{cases}$$

$$\mathbf{J}_{Aj}^{(i)} = \begin{cases} \mathbf{0} & \text{for a prismatic joint} \\ \mathbf{b}_{j-1} & \text{for a revolute joint} \end{cases}$$

$$(5-26)$$

where $\mathbf{r}_{0,ci}$ is the position vector of the centroid of link i referred to the base coordinate frame, and $\mathbf{b}_{j-1}$ is the 3×1 unit vector along joint axis $j-1$.

Substituting expressions (5-24) into equations (5-22) and (5-23) yields

$$T = \frac{1}{2}\sum_{i=1}^{n} \left(m_i \dot{\mathbf{q}}^T \mathbf{J}_L^{(i)T} \mathbf{J}_L^{(i)} \dot{\mathbf{q}} + \dot{\mathbf{q}}^T \mathbf{J}_A^{(i)T} \mathbf{I}_i \mathbf{J}_A^{(i)} \dot{\mathbf{q}} \right) = \frac{1}{2}\, \dot{\mathbf{q}}^T \mathbf{H} \dot{\mathbf{q}} \tag{5-27}$$

where $\mathbf{H}$ is the $n \times n$ matrix given by

$$\mathbf{H} = \sum_{i=1}^{n} \left(m_i \mathbf{J}_L^{(i)T} \mathbf{J}_L^{(i)} + \mathbf{J}_A^{(i)T} \mathbf{I}_i \mathbf{J}_A^{(i)} \right) \tag{5-28}$$

The matrix $\mathbf{H}$ incorporates all the mass properties of the whole arm linkage, as reflected to the joint axes, and is referred to as the *manipulator inertia tensor*[1]. Note the difference between the manipulator inertia tensor and the 3×3 inertia tensors of the individual arm links. The former is a composite inertia tensor including the latter as components. The manipulator inertia tensor, however, has properties similar to those of individual inertia tensors. As shown in equation (5-28), the manipulator inertia tensor is a symmetric matrix, as is the individual inertia tensor defined by equation (5-2). The quadratic form associated with the manipulator inertia tensor represents kinetic energy, so does the individual inertia tensor. Kinetic energy is always strictly positive unless the system is at rest. The manipulator inertia tensor of equation (5-28) is positive definite, so are the individual inertia tensors. Note, however, that the manipulator inertia tensor involves Jacobian matrices, which vary with arm configuration. Therefore the manipulator inertia tensor is *configuration-dependent* and represents the instantaneous composite mass properties of the whole arm linkage at the current arm configuration.

Let H_{ij} be the $[i,j]$ component of the manipulator inertia tensor $\mathbf{H}$, then we can rewrite equation (5-27) in a scalar form so that

$$T = \frac{1}{2}\sum_{i=1}^{n} \sum_{j=1}^{n} H_{ij}\, \dot{q}_i \dot{q}_j \tag{5-29}$$

---

[1] This standard terminology is an abbreviation of *manipulator inertia tensor matrix* : strictly speaking, $\mathbf{H}$ is a matrix based on the individual inertia tensors.

Note that H_{ij} is a function of $q_1, \cdots, q_n$.

5.2.3. Deriving Lagrange's Equations of Motion

In addition to the computation of the kinetic energy we need to find the potential energy U and generalized forces in order to derive Lagrange's equations of motion. Let **g** be the 3×1 vector representing the acceleration of gravity with reference to the base coordinate frame, which is an inertial reference frame. Then the potential energy stored in the whole arm linkage is given by

$$U = \sum_{i=1}^{n} m_i \mathbf{g}^T \mathbf{r}_{0,ci} \qquad (5-30)$$

where the position vector of the centroid C_i is dependent on the arm configuration. Thus the potential function is a function of $q_1, \cdots, q_n$.

Generalized forces account for all the forces and moments acting on the arm linkage except gravity forces and inertial forces. We consider the situation where actuators exert joint torques $\boldsymbol{\tau} = [\tau_1, \cdots, \tau_n]^T$ at individual joints and an external force and moment $\mathbf{F}_{ext}$ is applied at the arm's endpoint while in contact with the environment. Generalized forces can be obtained by computing the virtual work done by these forces. In equation (4-9), let us replace the endpoint force exerted by the manipulator by the negative external force $-\mathbf{F}_{ext}$. Then the virtual work is given by

$$\delta\text{Work} = \boldsymbol{\tau}^T \delta\mathbf{q} + \mathbf{F}_{ext}^T \delta\mathbf{p} = (\boldsymbol{\tau} + \mathbf{J}^T \mathbf{F}_{ext})^T \delta\mathbf{q} \qquad (5-31)$$

By comparing this expression with the one in terms of generalized forces $\mathbf{Q} = [Q_1, \cdots, Q_n]^T$, given by

$$\delta\text{Work} = \mathbf{Q}^T \delta\mathbf{q} \qquad (5-32)$$

we can identify the generalized forces as

$$\mathbf{Q} = \boldsymbol{\tau} + \mathbf{J}^T \mathbf{F}_{ext} \tag{5-33}$$

Using the total kinetic energy (5-29) and the total potential energy (5-30), we can now derive Lagrange's equations of motion. From equation (5-29), the first term in equation (5-21) is computed as

$$\frac{d}{dt}\left(\frac{\partial T}{\partial \dot{q}_i}\right) = \frac{d}{dt}\left(\sum_{j=1}^{n} H_{ij}\dot{q}_j\right) = \sum_{j=1}^{n} H_{ij}\ddot{q}_j + \sum_{j=1}^{n} \frac{dH_{ij}}{dt}\dot{q}_j \tag{5-34}$$

Note that H_{ij} is a function of $q_1, \cdots, q_n$, so that the time derivative of H_{ij} is given by

$$\frac{dH_{ij}}{dt} = \sum_{k=1}^{n} \frac{\partial H_{ij}}{\partial q_k}\frac{dq_k}{dt} = \sum_{k=1}^{n} \frac{\partial H_{ij}}{\partial q_k}\dot{q}_k \tag{5-35}$$

The second term in equation (5-21) includes the partial derivative of the kinetic energy, given by

$$\frac{\partial T}{\partial q_i} = \frac{\partial}{\partial q_i}\left(\frac{1}{2}\sum_{j=1}^{n}\sum_{k=1}^{n} H_{jk}\dot{q}_j\dot{q}_k\right) = \frac{1}{2}\sum_{j=1}^{n}\sum_{k=1}^{n} \frac{\partial H_{jk}}{\partial q_i}\dot{q}_j\dot{q}_k \tag{5-36}$$

since H_{jk} depends on q_i. The gravity term G_i is obtained by taking the partial derivative of the potential energy:

$$G_i = \frac{\partial U}{\partial q_i} = \sum_{j=1}^{n} m_j \mathbf{g}^T \frac{\partial \mathbf{r}_{0,cj}}{\partial q_i} = \sum_{j=1}^{n} m_j \mathbf{g}^T \mathbf{J}_{Li}^{(j)} \tag{5-37}$$

since the partial derivative of the position vector $\mathbf{r}_{0,cj}$ with respect to q_i is the same as the i-th column vector of the Jacobian matrix $\mathbf{J}_{Li}^{(j)}$ defined by equations (5-24)-(5-26). Substituting expressions (5-34) through (5-37) into (5-21) yields

$$\sum_{j=1}^{n} H_{ij}\ddot{q}_j + \sum_{j=1}^{n}\sum_{k=1}^{n} h_{ijk}\dot{q}_j\dot{q}_k + G_i = Q_i \qquad i = 1, \cdots, n \tag{5-38}$$

where

$$h_{ijk} = \frac{\partial H_{ij}}{\partial q_k} - \frac{1}{2}\frac{\partial H_{jk}}{\partial q_i} \tag{5-39}$$

and

$$G_i = \sum_{j=1}^{n} m_j \mathbf{g}^T \mathbf{J}_{L\,i}^{(j)} \tag{5-40}$$

The first term represents inertia torques, including interaction torques, while the second term accounts for the Coriolis and centrifugal effects, and the last term is the gravity torque. It is important to note that interactive inertia torques $H_{ij}\ddot{q}_j$ $(j \neq i)$ result from the off-diagonal elements of the manipulator inertia tensor and that the Coriolis and centrifugal torques $h_{ijk}\dot{q}_j\dot{q}_k$ arise because the manipulator inertia tensor is configuration dependent. Equation (5-38) is the same as equation (5-13) derived from Newton-Euler equations. Thus the Lagrangian formulation provides the closed-form dynamic equations directly.

Example 5-2

Let us derive closed-form dynamic equations for the two degree-of-freedom planar manipulator shown in Figure 5-2, using Lagrange's equations of motion.

We begin by computing the manipulator inertia tensor $\mathbf{H}$. From equation (5-10), velocities of the centroids C_1 and C_2 can be written as

$$\mathbf{v}_{c1} = \begin{bmatrix} -l_{c1}\sin(\theta_1) & 0 \\ l_{c1}\cos(\theta_1) & 0 \end{bmatrix}\dot{\mathbf{q}}$$

$$\tag{5-41}$$

$$\mathbf{v}_{c2} = \begin{bmatrix} -l_1\sin(\theta_1) - l_{c2}\sin(\theta_1+\theta_2) & -l_{c2}\sin(\theta_1+\theta_2) \\ l_1\cos(\theta_1) + l_{c2}\cos(\theta_1+\theta_2) & l_{c2}\cos(\theta_1+\theta_2) \end{bmatrix}\dot{\mathbf{q}}$$

The above 2×2 matrices are the Jacobian matrices $\mathbf{J}_L^{(i)}$ of equation (5-24). The angular velocities are associated with the Jacobian matrices $\mathbf{J}_A^{(i)}$, which are 1×2 row-vectors in this planar case:

$$\omega_1 = \dot{\theta}_1 = [1 \quad 0]\dot{\mathbf{q}}$$

$$\omega_2 = \dot{\theta}_1 + \dot{\theta}_2 = [1 \quad 1]\dot{\mathbf{q}} \tag{5-42}$$

Substituting the above expressions into equation (5-28), we obtain the manipulator inertia tensor

$$\tag{5-43}$$

$$\mathbf{H} = \begin{bmatrix} m_1 l_{c1}^2 + I_1 + m_2(l_1^2 + l_{c2}^2 + 2\,l_1 l_{c2} \cos \theta_2) + I_2 & m_2 \, l_1 l_{c2} \cos \theta_2 + m_2 \, l_{c2}^2 + I_2 \\ m_2 \, l_1 l_{c2} \cos \theta_2 + m_2 \, l_{c2}^2 + I_2 & m_2 \, l_{c2}^2 + I_2 \end{bmatrix}$$

The components of the above inertia tensor are the coefficients of the first term of equation (5-38). The second term is determined by substituting equation (5-43) into equation (5-39).

$$\begin{bmatrix} h_{111} = 0 \;,\; h_{122} = -m_2 \, l_1 l_{c2} \sin \theta_2 \;,\; h_{112} + h_{121} = -2\,m_2 \, l_1 l_{c2} \sin \theta_2 \\ h_{211} = m_2 \, l_1 l_{c2} \sin \theta_2 \;,\; h_{222} = 0 \;,\; h_{212} + h_{221} = 0 \end{bmatrix} \tag{5-44}$$

The third term in Lagrange's equations of motion, i.e., the gravity term, is derived from equation (5-40) using the Jacobian matrices in equation (5-41) :

$$G_1 = \mathbf{g}^T [m_1 \mathbf{J}_{L1}^{(1)} + m_2 \mathbf{J}_{L1}^{(2)}]$$

$$G_2 = \mathbf{g}^T [m_1 \mathbf{J}_{L2}^{(1)} + m_2 \mathbf{J}_{L2}^{(2)}] \tag{5-45}$$

Substituting equations (5-43), (5-44) and (5-45) into equation (5-38) yields

$$H_{11}\ddot{\theta}_1 + H_{12}\ddot{\theta}_2 + h_{122}\dot{\theta}_2^2 + (h_{112}+h_{121})\dot{\theta}_1\dot{\theta}_2 + G_1 = \tau_1$$

$$H_{22}\ddot{\theta}_2 + H_{12}\ddot{\theta}_1 + h_{211}\dot{\theta}_1^2 + G_2 = \tau_2 \tag{5-46}$$

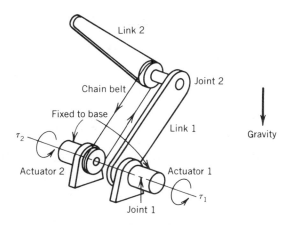

Figure 5-7 :   Remotely driven two d.o.f. planar manipulator.

Note that, since no external force acts on the endpoint, the generalized forces coincide with the joint torques, as shown in equation (5-33). Equation (5-46) is the same as equation (5-11), which was derived from the Newton-Euler equations.   △△△

Example 5-3

Figure 5-7 shows a planar manipulator whose arm links have the same mass properties as those of the manipulator of Figure 5-2. The actuators and transmissions, however, are different. The second actuator, driving joint 2, is now located at the base, and the output torque is transmitted to joint 2 through a chain drive mechanism. Since the actuator is fixed to the base link, its reaction torque acts on the base link, while in Figure 5-2 the reaction torque of the second actuator acts on link 1. The first actuator, on the other hand, is the same for the two manipulators. Let us find Lagrange's equations of motion for this remotely driven manipulator.

The manipulator inertia tensor and the potential function are the same as for the manipulator of Figure 5-2. Let us investigate the virtual work done by the generalized forces. Letting τ_1^* and τ_2^* be the torques exerted by the first and the second actuators, respectively, the virtual work done by these torques is

$$\delta\text{Work} = \overset{*}{\tau}_1 \, \delta\theta_1 + \overset{*}{\tau}_2 \, (\delta\theta_1 + \delta\theta_2)$$

$$= (\overset{*}{\tau}_1 + \overset{*}{\tau}_2) \, \delta\theta_1 + \overset{*}{\tau}_2 \, \delta\theta_2 \qquad\qquad (5-47)$$

Comparing the above expression with (5-32):

$$\delta\text{Work} = \mathbf{Q}^T \, \delta\mathbf{q} = Q_1 \delta q_1 + Q_2 \delta q_2$$

where $\delta q_1 = \delta\theta_1$ and $\delta q_2 = \delta\theta_2$, we find that the generalized forces are

$$Q_1 = \overset{*}{\tau}_1 + \overset{*}{\tau}_2 \qquad\qquad Q_2 = \overset{*}{\tau}_2 \qquad\qquad (5-48)$$

Replacing τ_1 and τ_2 in equation (5-46) by Q_1 and Q_2, respectively, we obtain the dynamic equations of the remotely driven manipulator. $\Delta\Delta\Delta$

5.2.4. Transformations of Generalized Coordinates

In the previous section, we used joint displacements as a complete set of independent generalized coordinates to describe Lagrange's equations of motion. However, any complete set of independent generalized coordinates can be used. It is a significant feature of the Lagrangian formulation that we can employ any convenient coordinates to describe the system. Also, in the Lagrangian formulation, coordinate transformations can be performed in a simple and systematic manner.

As before, let $\mathbf{q} = [q_1, \cdots, q_n]^T$ be the vector of joint coordinates, which represents a complete and independent set of generalized coordinates. We now assume that there exists another set of complete and independent generalized coordinates, $\mathbf{p} = [p_1, \cdots, p_n]^T$, that satisfy the following differential relationship with $\mathbf{q}$:

$$d\mathbf{p} = \mathbf{J}d\mathbf{q} \qquad\qquad (5-49)$$

The Jacobian matrix $\mathbf{J}$ is assumed to be a non-singular square matrix within a specified region in $\mathbf{q}$-coordinates. Let us derive Lagrange's equations of motion in $\mathbf{p}$-coordinates from the ones expressed in **q**-coordinates. To this end, we must find the transforms of the manipulator inertia tensor **H**, the Coriolis and centrifugal coefficients h_{ijk}, and the derivatives G_i of the potential function U.

From equations (5-27) and (5-49), the kinetic energy can be expressed in terms of $\dot{\mathbf{p}}$ as

$$T = \frac{1}{2}\dot{\mathbf{p}}^T \mathbf{H}^* \dot{\mathbf{p}} \tag{5-50}$$

where

$$\mathbf{H}^* = (\mathbf{J}^{-1})^T \mathbf{H} \mathbf{J}^{-1} \tag{5-51}$$

The matrix $\mathbf{H}^*$ represents the manipulator inertia tensor referred to **p**-coordinates. The transformation of inertia tensors is thus given by equation (5-51). The first term of Lagrange's equations of motion is determined by the new manipulator inertia tensor $\mathbf{H}^*$. The Coriolis and centrifugal terms are derived by differentiating $\mathbf{H}^*$ as in (5-39) :

$$h_{ijk}^* = \frac{\partial H_{ij}^*}{\partial p_k} - \frac{1}{2}\frac{\partial H_{jk}^*}{\partial p_i} = \sum_{l=1}^{n}\left(\frac{\partial H_{ij}^*}{\partial q_l}\hat{J}_{lk} - \frac{1}{2}\frac{\partial H_{jk}^*}{\partial q_l}\hat{J}_{li}\right) \tag{5-52}$$

where $\hat{J}_{lk}$ is the $[l, k]$ element of the inverse Jacobian matrix $\mathbf{J}^{-1}$. Gravity terms in **p**-coordinates, G_i^*, are derived from differentiating the potential function U in terms of **p**. From (5-37) and (5-49), we get

$$G_i^* = \frac{\partial U}{\partial p_i} = \sum_{j=1}^{n}\frac{\partial U}{\partial q_j}\frac{\partial q_j}{\partial p_i} = \sum_{j=1}^{n}G_j\hat{J}_{ji} \tag{5-53}$$

or in vector form

$$\mathbf{G}^* = (\mathbf{J}^{-1})^T \mathbf{G} \tag{5-54}$$

where $\mathbf{G}^*$ and $\mathbf{G}$ are $n \times 1$ vectors of components G_i^* and G_i ($i = 1, \cdots, n$), respectively.

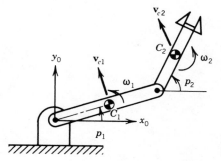

Figure 5-8 :   Representation of arm configuration using absolute angles p_1 and p_2 .

Finally let $\mathbf{Q}^* = [Q_1^*, \;\cdots\; ,Q_n^*]^T$ be the generalized forces in $\mathbf{p}$-coordinates. The principle of virtual work yields

$$\delta\text{Work} = \mathbf{Q}^T \delta\mathbf{q} = \mathbf{Q}^T \mathbf{J}^{-1} \delta\mathbf{p} = \mathbf{Q}^* \delta\mathbf{p} \qquad (5-55)$$

and therefore

$$\mathbf{Q}^* = (\mathbf{J}^{-1})^T \mathbf{Q} \qquad (5-56)$$

$\triangle\triangle\triangle$

Example 5-4

Consider again the two degree-of-freedom planar manipulator of Figure 5-7, where the second joint is remotely driven by the actuator fixed to the base. We now use the angles p_1 and p_2 shown in Figure 5-8 as generalized coordinates. The new coordinates represent the absolute angles of the two links measured from the base line (the x_0 axis), whereas the joint displacements θ_1 and θ_2 represent the relative angles between adjacent links. The two angles p_1 and p_2 are independent variables, and furthermore determine the arm configuration completely. Therefore, they can indeed be used as generalized coordinates. Let us derive the equations of motions in the p_1, p_2 coordinates.

We first obtain the manipulator inertia tensor in $\mathbf{p}$-coordinates. The total kinetic energy stored in the two links is given by

$$T = \frac{1}{2} m_1 |\mathbf{v}_{c1}|^2 + \frac{1}{2} I_1 (\omega_1)^2 + \frac{1}{2} m_2 |\mathbf{v}_{c2}|^2 + \frac{1}{2} I_2 (\omega_2)^2 \tag{5-57}$$

where

$$|\mathbf{v}_{c1}|^2 = l_{c1}^2 \, \dot{p}_1^2$$

$$|\mathbf{v}_{c2}|^2 = l_1^2 \, \dot{p}_1^2 + l_{c2}^2 \, \dot{p}_2^2 + 2 \, l_1 l_{c2} \, \dot{p}_1 \dot{p}_2 \cos (p_2 - p_1)$$

$$\omega_1 = \dot{p}_1 \qquad\qquad \omega_2 = \dot{p}_2 \tag{5-58}$$

Rewriting the total kinetic energy in the quadratic form (5-27), we find the components H_{ij}^{*} of the manipulator inertia tensor in **p**-coordinates:

$$H_{11}^{*} = m_1 \, l_{c1}^2 + I_1 + m_2 \, l_1^2$$

$$H_{22}^{*} = m_2 \, l_{c2}^2 + I_2 \tag{5-59}$$

$$H_{12}^{*} = m_2 \, l_1 l_{c2} \cos (p_2 - p_1)$$

Let us now show that the same result can be obtained by the coordinate transformation of manipulator inertia tensors given by equation (5-51). From the figure, the relationship between the two sets of generalized coordinates is given by

$$p_1 = \theta_1 \qquad\qquad p_2 = \theta_1 + \theta_2 \tag{5-60}$$

The inverse manipulator Jacobian associated with this coordinate transformation is thus given by

$$\mathbf{J}^{-1} = \begin{bmatrix} 1 & 0 \\ -1 & 1 \end{bmatrix} \tag{5-61}$$

Substituting (5-61) into (5-51) yields

$$H_{11}^{*} = H_{11} + H_{22} - 2 \, H_{12}$$

$$H_{12}^{*} = H_{12} - H_{22} \tag{5-62}$$

$$H_{22}^{*} = H_{22}$$

where H_{ij} is the $[i\,,j]$ element of **H** that was obtained in equation (5-12). Substituting (5-12) into (5-62), we obtain the same result as (5-59).

From equations (5-56), (5-61), and (5-48), the transformation of generalized forces is given by

$$Q_1^* = Q_1 - Q_2 = \tau_1^*$$
$$Q_2^* = Q_2 = \tau_2^* \tag{5-63}$$

Similarly, the gravity terms G_1 and G_2 are transformed into G_1^* and G_2^*. Lagrange's equations of motion in **p**-coordinates are then given by

$$H_{11}^*\,\ddot{p}_1 + H_{12}^*\,\ddot{p}_2 + \frac{\partial H_{12}^*}{\partial p_2}\,\dot{p}_2^2 + G_1^* = \tau_1^*$$

$$H_{22}^*\,\ddot{p}_2 + H_{12}^*\,\ddot{p}_1 + \frac{\partial H_{12}^*}{\partial p_1}\,\dot{p}_1^2 + G_2^* = \tau_2^* \tag{5-64}$$

Note that in **p**-coordinates the diagonal elements of H^* are configuration-invariant and that the Coriolis torque, which is proportional to the product $\dot{p}_1\,\dot{p}_2$, does not appear. This can be easily understood. In **q**-coordinates, the motion of link 2 is represented relative to link 1, which rotates with an angular velocity $\dot{q}_1$. In other words, the motion of link 2 is represented relative to the moving coordinate attached to link 1. Therefore, a Coriolis torque arises when link 2 moves while link 1 rotates. In **p**-coordinates, however, the rotation of link 2 is represented with reference to the base frame and is independent of link 1, hence there is no Coriolis effect. Thus, the equations of motion can be simplified by selecting appropriate generalized coordinates.   △△△

Example 5-5

In the kinematic and static analysis of a manipulator arm, we are concerned with the motion of the end-effector, because of its direct influence upon the task to be accomplished.

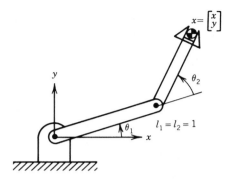

Figure 5-9 :   Representation of arm configuration
using endpoint coordinates x and y.

Similarly, let us now consider the dynamic equations for the end-effector motion, using endpoint coordinates.

Consider the two degree-of-freedom manipulator of Figure 5-9. We assume that the range of joint 2 is limited within $0 < \theta_2 < \pi$. Under this condition, the solution to the kinematic equation is unique: given arbitrary endpoint coordinates x and y within the reachable range, joint displacements θ_1 and θ_2 are uniquely determined. Therefore, we can use endpoint coordinates x and y as a complete and independent set of generalized coordinates, in the same way as joint coordinates. When the second joint is limited to the range $0 < \theta_2 < \pi$, the Jacobian matrix associated with the endpoint motion remains non-singular, as shown in Example 3-2. From (3-34), the inverse Jacobian is given by

$$
\mathbf{J}^{-1}(\theta_1, \theta_2) \;=\; \frac{1}{\sin\theta_2}
\begin{bmatrix}
\sin(\theta_1 + \theta_2) & -\cos(\theta_1 + \theta_2) \\
-\sin(\theta_1) - \sin(\theta_1 + \theta_2) & \cos(\theta_1) + \cos(\theta_1 + \theta_2)
\end{bmatrix}
\tag{5-65}
$$

If we denote by $\mathbf{H}$ the manipulator inertia tensor in joint coordinates, the manipulator inertia tensor referred to endpoint coordinates is given by $\mathbf{H}^{*} = (\mathbf{J}^{-1})^{T}\mathbf{H}\mathbf{J}^{-1}$, which is a function of θ_1 and θ_2. The equations of motion with respect to the endpoint motion are then derived from $\mathbf{H}^{*}$ above and equations (5-52), (5-54) and (5-56).   △△△

5.3. Inverse Dynamics

5.3.1. Introduction

The closed-form dynamic equations derived in the previous sections govern the dynamic responses of a manipulator arm to the input joint torques generated by the actuators. This dynamic process can be illustrated by the block diagram of Figure 5-10, where the inputs are joint torques $\tau_1(t), \cdots, \tau_n(t)$, and the outputs are generalized coordinates, typically joint displacements $q_1(t), \cdots, q_n(t)$. As discussed in previous chapters, inverse problems are important to robot control and programming, since they allow one to find the appropriate inputs necessary for producing the desired outputs.

In this section, we discuss the inverse dynamics process, shown in the block diagram at the bottom of Figure 5-10. The inputs are the desired trajectories, described as time functions

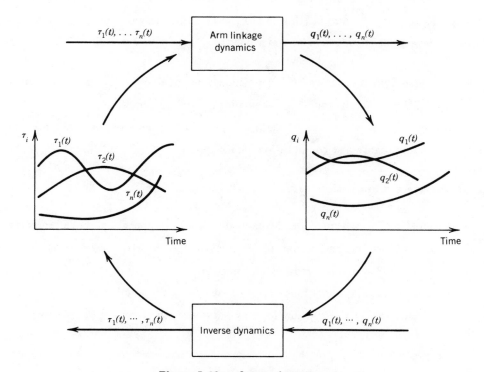

Figure 5-10 :   Inverse dynamics.

$q_1(t)$ through $q_n(t)$. The outputs are the joint torques to be applied at each instant by the actuators in order to follow the specified trajectories, and are obtained by evaluating the right-hand side of the closed-form dynamic equations

$$\tau_i = \sum_{j=1}^{n} H_{ij}\, \ddot{q}_j + \sum_{j=1}^{n}\sum_{k=1}^{n} h_{ijk}\dot{q}_j\,\dot{q}_k + G_i \qquad\qquad i=1,\cdots,n$$

using the specified trajectory data. At each instant we compute joint velocities $\dot{q}_j$ and joint accelerations $\ddot{q}_j$ from the given time functions, and then substitute them to the right-hand side of the above equation. It must be noted that the coefficients, H_{ij}, h_{ijk}, and G_i are all configuration-dependent. When all the coefficients need to be computed, the total amount of computation becomes extremely large. As we have seen in equations (5-28) and (5-39), the computation required for the first and the second terms of Lagrange's equations increases quite rapidly as the number of degrees of freedom n increases; the number of multiplications required for the first term is approximately proportional to n^3, while that required for the third term is proportional to n^4. For a six degree-of-freedom manipulator arm, we end up with 66271 multiplications for each data point (Hollerbach, 1981). Thus, the extremely heavy computation load is a bottleneck for the inverse dynamics.

The inverse dynamics approach is particularly important for control, since it allows us to compensate for the highly coupled and nonlinear arm dynamics, as discussed in the next chapter. However, we need to cope with the computational complexity in real time. Thus, in this section, we investigate fast computation algorithms .

5.3.2. Recursive Computation

Two efficient algorithms for inverse dynamics computation have recently been developed. One is based on the Lagrangian formulation and the other is based on the Newton-Euler formulation. Both methods reduce the computational complexity from $O(n^4)$ to $O(n)$, so that the required number of operations varies linearly with the number of degrees of freedom. This reduction is particularly significant for manipulators with many degrees of freedom.

The key concept of both methods is to formulate dynamic equations in a *recursive* form, so that the computation can be accomplished from one link of a manipulator arm to another.

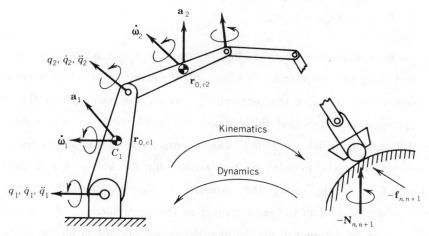

Figure 5-11 :   Recursive computation of kinematic and dynamic equations.

Figure 5-11 illustrates the outline of the recursive computation algorithm based on the Newton-Euler formulation. The algorithm can be applied to any manipulator arm with an open kinematic chain structure.

The first phase of the recursive Newton-Euler formulation is to determine all the kinematic variables that are needed for evaluating the Newton-Euler equations. These include the linear and angular velocities and accelerations of each link member involved in the serial linkage. The algorithm starts with the first link. Given the joint displacement q_1, and the joint velocity and acceleration $\dot{q}_1$ and $\ddot{q}_1$, the linear and angular velocities and accelerations of the centroid C_1 are determined. Then, using the velocities and accelerations of the first link, denoted by $\mathbf{v}_{c1}$, $\boldsymbol{\omega}_1$, $\mathbf{a}_{c1}$ and $\dot{\boldsymbol{\omega}}_1$, we compute the velocity and acceleration of the second link with the data specified for joint 2, namely q_2, $\dot{q}_2$, and $\ddot{q}_2$. This procedure is repeated until all the centroidal velocities and accelerations, as well as the angular velocities and accelerations, are determined for all the links involved.

The second phase of the recursive formulation is to evaluate Newton-Euler equations with the computed kinematic variables to determine the joint torques. We now proceed with the recursive computation starting from the last link back to the proximal links. Let us recall the force/momentum relationship given by equation (5-1). We can rewrite the equation for link n as

$$\mathbf{f}_{n-1,n} \,=\, \mathbf{f}_{n,n+1} - m_n \mathbf{g} + m_n \mathbf{a}_{cn} \tag{5-66}$$

where $\mathbf{f}_{n-1,n}$ is the coupling force between links $n-1$ and n, $\mathbf{f}_{n,n+1}$ is the linear endpoint force that is specified along with trajectories to follow, $\mathbf{g}$ is the gravitational acceleration vector, and $\mathbf{a}_{cn}$ is the acceleration vector of the centroid C_n, which was computed in the first phase. From this expression, it follows that the unknown coupling force $\mathbf{f}_{n-1,n}$ can be determined by evaluating the right-hand side of (5-66), which consists of known or specified variables. Similarly, we can write the force/momentum relationship for link $n-1$ and determine the coupling force $\mathbf{f}_{n-2,n-1}$ by using the variables previously obtained. Moment/angular momentum can be evaluated in the same manner as the linear forces, and thus the coupling moments $\mathbf{N}_{i-1,i}$ can be determined one by one. Hence, we can obtain all the coupling forces and moments recursively, by evaluating the dynamic equations from the last link back to the first link.

To summarize, the recursive procedure can be formulated as

$$\mathbf{f}_{i-1,i} = \mathbf{f}_{i,i+1} - m_i \mathbf{g} + m_i \mathbf{a}_{ci} \tag{5-67}$$

$$\mathbf{N}_{i-1,i} = \mathbf{N}_{i,i+1} - \mathbf{r}_{i,ci} \times \mathbf{f}_{i,i+1} + \mathbf{r}_{i-1,ci} \times \mathbf{f}_{i-1,i} + \mathbf{I}_i \dot{\boldsymbol{\omega}}_i + \boldsymbol{\omega}_i \times (\mathbf{I}_i \boldsymbol{\omega}_i) \tag{5-68}$$

This procedure is repeated until the link number i reaches $i = 1$. Once the coupling force and moment of each joint are determined, the joint torque can be computed from (4-4) or (4-5), depending on the type of joint.

5.3.3. Moving Coordinates

In the first phase of the computation, we need to find the velocity and acceleration of link i, given the motion of the previous link and the specified motion of link i relative to link $i-1$. To solve this problem, we need to analyze relative motions defined in a moving coordinate frame. In this section we derive basic results about moving coordinates, and then apply these results to the recursive computation algorithm.

Let us analyze the motion of a vector represented with reference to a moving coordinate frame, as shown in Figure 5-12. The coordinate frame $O_0 - x_0 y_0 z_0$ is fixed to the ground, while

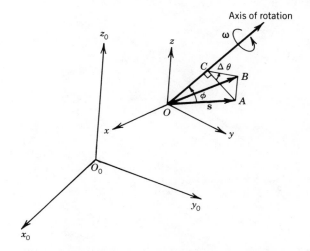

Figure 5-12 :   Time rate of change of a vector fixed to a rotating coordinate frame.

$O-xyz$ is rotating with an angular velocity $\boldsymbol{\omega}$. The origin O itself is assumed to be stationary in the figure. An arbitrary vector $\mathbf{s}$ is fixed to $O-xyz$, and thus moves with the rotating coordinate frame. Let us first compute the time rate of change of vector $\mathbf{s}$ as viewed from the fixed frame:

$$\frac{d\mathbf{s}}{dt}\Big|_{\text{fixed}} \tag{5-69}$$

Consider a short time interval Δt. The moving coordinate frame rotates $|\Delta\theta| = |\boldsymbol{\omega}|\Delta t$ about the axis of rotation as shown in the figure. Accordingly, the vector $\mathbf{s}$ moves from point A to point B. Let ϕ be the angle $\angle AOC$ in the figure, then the magnitude of the change in vector $\mathbf{s}$ is

$$\overline{AB} = \overline{AC}\,|\Delta\theta| = \overline{AO}\,\sin(\phi)\,|\boldsymbol{\omega}|\Delta t = |\mathbf{s}|\,|\boldsymbol{\omega}|\sin(\phi)\times\Delta t \tag{5-70}$$

The vector $\overrightarrow{A\ B}$ is perpendicular to both the axis of rotation and vector $\mathbf{s}$, hence is parallel to the vector product $\boldsymbol{\omega}\times\mathbf{s}$. Thus, the time rate of change of the vector $\mathbf{s}$ as viewed from the fixed coordinate frame is given by

$$\frac{d\mathbf{s}}{dt}\Big|_{\text{fixed}} = \boldsymbol{\omega}\times\mathbf{s} \tag{5-71}$$

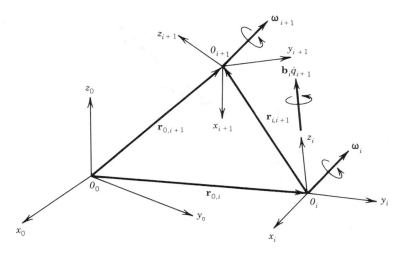

Figure 5-13 :   Motion relative to a moving coordinate frame.

Figure 5-13 shows coordinate frames fixed to the base, link i , and link $i+1$, denoted respectively by $O_0-x_0y_0z_0$, $O_i-x_iy_iz_i$ and $O_{i+1}-x_{i+1}y_{i+1}z_{i+1}$. From the figure,

$$\mathbf{r}_{0,i+1} = \mathbf{r}_{0,i} + \mathbf{r}_{i,i+1} \tag{5-72}$$

We derive the time rate of change of the right-hand side of (5-72) when the frame $O_i-x_iy_iz_i$ is rotated at an angular velocity $\boldsymbol{\omega}_i$. The time rate of change of each term in equation (5-72), when viewed from the base coordinate frame, is

$$\frac{d\mathbf{r}_{0,i+1}}{dt}\Big|_{\text{fixed}} = \frac{d\mathbf{r}_{0,i}}{dt}\Big|_{\text{fixed}} + \frac{d\mathbf{r}_{i,i+1}}{dt}\Big|_{\text{fixed}} \tag{5-73}$$

The suffix "*fixed* " is to indicate that the time rate of change is viewed from the fixed coordinate frame.   Vectors $\mathbf{r}_{0,i+1}$ and $\mathbf{r}_{0,i}$ are defined with reference to the base frame, and their time derivatives, referred to the base frame, are denoted by $\mathbf{v}_{i+1}$ and $\mathbf{v}_i$.   However, vector $\mathbf{r}_{i,i+1}$ represents the relative displacement with respect to the moving coordinate frame. Let $\mathbf{n}$, $\mathbf{t}$, and $\mathbf{b}$ be unit vectors along the coordinate axes of the moving frame $O_i-x_iy_iz_i$ at the instant shown.   Also, let x, y, and z be the components of $\mathbf{r}_{i,i+1}$ with respect to the moving frame, then

$$\frac{d\mathbf{r}_{i,i+1}}{dt}\Big|_{\text{fixed}} = \frac{d}{dt}\,[x\,\mathbf{n} + y\,\mathbf{t} + z\,\mathbf{b}]$$

$$= \left(\frac{dx}{dt}\,\mathbf{n} + \frac{dy}{dt}\,\mathbf{t} + \frac{dz}{dt}\,\mathbf{b}\right) + \left(x\,\frac{d\mathbf{n}}{dt} + y\,\frac{d\mathbf{t}}{dt} + z\,\frac{d\mathbf{b}}{dt}\right) \qquad (5-74)$$

The first term may be interpreted as the velocity contribution due to the motion of point O_{i+1} relative to O_i. Let us denote this term by

$$\frac{d\mathbf{r}_{i,i+1}}{dt}\Big|_{\text{rel.}} \qquad (5-75)$$

to indicate that the time rate of change is viewed from the moving frame. The second term in equation (5-74) may be interpreted as the velocity contribution induced by the rotation of the moving frame. Since vectors $\mathbf{n}$, $\mathbf{t}$, and $\mathbf{b}$ are fixed to the moving frame and thus move with it, their time rates of change as viewed from the base frame are given by equation (5-71). Then, from equations (5-71), (5-73), (5-74), and (5-75), we obtain

$$\mathbf{v}_{i+1} = \mathbf{v}_i + \frac{d\mathbf{r}_{i,i+1}}{dt}\Big|_{\text{rel.}} + \boldsymbol{\omega}_i \times \mathbf{r}_{i,i+1} \qquad (5-76)$$

Although the vector $\mathbf{r}_{i,i+1}$ was defined to be a position vector, in the derivation of the second and third terms of equation (5-76), the result does not depend on the specific meaning of the vector. The same derivation can be applied to any vector. In general, the time rate of change of an arbitrary vector that moves relatively to a rotating coordinate frame is computed with the differential operator symbolically denoted by

$$\frac{d}{dt}\Big|_{\text{fixed}} = \frac{d}{dt}\Big|_{\text{rel.}} + \boldsymbol{\omega} \times \qquad (5-77)$$

We can also obtain the second derivative of $\mathbf{r}_{0,i+1}$ from equation (5-76) by applying the differential operator (5-77) repeatedly:

$$\frac{d\mathbf{v}_{i+1}}{dt}\Big|_{\text{fixed}} = \frac{d\mathbf{v}_i}{dt}\Big|_{\text{fixed}} + \frac{d}{dt}\Big|_{\text{fixed}}\left(\frac{d\mathbf{r}_{i,i+1}}{dt}\Big|_{\text{rel.}}\right) + \frac{d\boldsymbol{\omega}_i}{dt}\Big|_{\text{fixed}} \times \mathbf{r}_{i,i+1} + \boldsymbol{\omega}_i \times \frac{d\mathbf{r}_{i,i+1}}{dt}\Big|_{\text{fixed}}$$

$$(5-78)$$

The left-hand side and the first term on the right-hand side represent, respectively, the accelerations of links $i+1$ and i, referred to the base frame. We denote them by $\mathbf{a}_{i+1}$ and $\mathbf{a}_i$, and apply again the differential operator to the other terms. Then,

$$\mathbf{a}_{i+1} = \mathbf{a}_i + \frac{d^2 \mathbf{r}_{i,i+1}}{dt^2}\Big|_{\text{rel.}} + \dot{\boldsymbol{\omega}}_i \times \mathbf{r}_{i,i+1} + 2\boldsymbol{\omega}_i \times \frac{d\mathbf{r}_{i,i+1}}{dt}\Big|_{\text{rel.}} + \boldsymbol{\omega}_i \times (\boldsymbol{\omega}_i \times \mathbf{r}_{i,i+1})$$

$$(5-79)$$

The second term on the right-hand side represents the relative acceleration viewed from the moving frame, the third term is the contribution due to the angular acceleration of the moving frame, the fourth term is the Coriolis acceleration, and the last term is the centrifugal acceleration due to the rotation of the moving frame.

5.3.4. Luh-Walker-Paul's Algorithm

On the basis of the kinematic analysis on the moving coordinate frame, we now formulate the recursive computation algorithm of Newton-Euler dynamic equations. The algorithm was originally developed by (Luh, Walker and Paul, 1980-a).

The first phase consists of kinematic computations. We derive different recursive equations depending on the type of joint (prismatic or revolute). When joint $i+1$ is prismatic, the angular velocity and acceleration of link $i+1$ are the same as those of the previous link:

$$\boldsymbol{\omega}_{i+1} = \boldsymbol{\omega}_i \qquad\qquad (5-80)$$

$$\dot{\boldsymbol{\omega}}_{i+1} = \dot{\boldsymbol{\omega}}_i \qquad\qquad (5-81)$$

On the other hand, if joint $i+1$ is revolute, the frame $i+1$ is rotated at an angular velocity $\dot{q}_{i+1}\mathbf{b}_i$ and with an angular acceleration $\ddot{q}_{i+1}\mathbf{b}_i$ about the z_i axis of the moving coordinate frame attached to link i. The angular velocity of link $i+1$ referred to the base frame is then given by

$$\boldsymbol{\omega}_{i+1} = \boldsymbol{\omega}_i + \dot{q}_{i+1}\mathbf{b}_i \qquad\qquad (5-82)$$

The recursive equation for angular acceleration can be obtained by simply taking the time derivative of both sides. Note, however, that the second term is defined as a vector relative to the moving coordinate frame. Hence the differential operator (5-77) must be employed in order to obtain the time derivative viewed from the base frame:

$$\dot{\omega}_{i+1} = \dot{\omega}_i + \ddot{q}_{i+1}\mathbf{b}_i + \omega_i \times \dot{q}_i\mathbf{b}_i \qquad (5\text{--}83)$$

The recursive equations for linear velocities and accelerations are derived from equations (5-76) and (5-79). The second terms in both equations are caused by the motion of joint $i+1$ relative to link i. If joint $i+1$ is prismatic,

$$\frac{d\mathbf{r}_{i,i+1}}{dt}\Big|_{\text{rel.}} = \dot{q}_{i+1}\mathbf{b}_i \qquad (5\text{--}84)$$

$$\frac{d^2\mathbf{r}_{i,i+1}}{dt^2}\Big|_{\text{rel.}} = \ddot{q}_{i+1}\mathbf{b}_i \qquad (5\text{--}85)$$

Substituting (5-84) and (5-85) into (5-76) and (5-79) then yields

$$\mathbf{v}_{i+1} = \mathbf{v}_i + \dot{q}_{i+1}\mathbf{b}_i + \omega_i \times \mathbf{r}_{i,i+1} \qquad (5\text{--}86)$$

$$\mathbf{a}_{i+1} = \mathbf{a}_i + \ddot{q}_{i+1}\mathbf{b}_i + \dot{\omega}_i \times \mathbf{r}_{i,i+1} + 2\omega_i \times \dot{q}_{i+1}\mathbf{b}_i + \omega_i \times (\omega_i \times \mathbf{r}_{i,i+1}) \qquad (5\text{--}87)$$

If joint $i+1$ is revolute,

$$\frac{d\mathbf{r}_{i,i+1}}{dt}\Big|_{\text{rel.}} = \dot{q}_{i+1}\mathbf{b}_i \times \mathbf{r}_{i,i+1} \qquad (5\text{--}88)$$

$$\frac{d^2\mathbf{r}_{i,i+1}}{dt^2}\Big|_{\text{rel.}} = \ddot{q}_{i+1}\mathbf{b}_i \times \mathbf{r}_{i,i+1} + \dot{q}_{i+1}\mathbf{b}_i \times (\dot{q}_{i+1}\mathbf{b}_i \times \mathbf{r}_{i,i+1}) \qquad (5\text{--}89)$$

Substituting (5-82) and (5-88) into (5-76), we get

$$\mathbf{v}_{i+1} = \mathbf{v}_i + \omega_{i+1} \times \mathbf{r}_{i,i+1} \qquad (5\text{--}90)$$

Further, substituting (5-83) and (5-89) into (5-79) and using the identity of vector triple products, i.e., $(\mathbf{a} \times \mathbf{b}) \times \mathbf{c} = (\mathbf{a}^T \mathbf{c})\mathbf{b} - (\mathbf{b}^T \mathbf{c})\mathbf{a}$ and $\mathbf{a} \times (\mathbf{b} \times \mathbf{c}) = (\mathbf{a}^T \mathbf{c})\mathbf{b} - (\mathbf{a}^T \mathbf{b})\mathbf{c}$, we

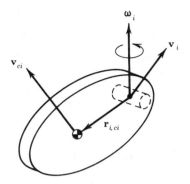

Figure 5-14 :   Centroid velocity and joint velocity.

obtain

$$\mathbf{a}_{i+1} = \mathbf{a}_i + \dot{\omega}_{i+1} \times \mathbf{r}_{i,i+1} + \omega_{i+1} \times (\omega_{i+1} \times \mathbf{r}_{i,i+1}) \qquad (5-91)$$

The Newton-Euler equations are expressed in terms of centroidal accelerations, whereas the recursive formulation is expressed with respect to the origin of the coordinate frame attached to each link.   Therefore, in order to evaluate (5-67), we need to transform all variables to centroidal variables.   This is illustrated in Figure 5-14, where $\mathbf{v}_i$ and ω_i are, respectively, the velocity at the origin of the coordinate frame attached to link i, and the angular velocity of the link.   The centroidal velocity is then given by

$$\mathbf{v}_{ci} = \mathbf{v}_i + \omega_i \times \mathbf{r}_{i,ci} \qquad (5-92)$$

Thus, by applying the differential operator (5-77) to expression (5-92), we find the centroidal acceleration:

$$\dot{\mathbf{v}}_{ci} = \dot{\mathbf{v}}_i + \dot{\omega}_i \times \mathbf{r}_{i,ci} + \omega_i \times (\omega_i \times \mathbf{r}_{i,ci}) \qquad (5-93)$$

Finally, we discuss the angular momentum involved in Euler's equation (5-68).   As mentioned before, the inertia tensor $\mathbf{I}_i$ varies depending on the orientation of the link.   Let $\mathbf{R}_i^0$ be the 3×3 rotation matrix associated with the coordinate transformation from frame i to the base frame, and $\overline{\mathbf{I}_i}$ be the inertia tensor expressed in the coordinate frame fixed to the link itself.   The inertia tensor $\overline{\mathbf{I}_i}$ is then given by

$$\mathbf{I}_i = \mathbf{R}_i^0 \, \overline{\mathbf{I}}_i \, (\mathbf{R}_i^0)^T \qquad\qquad (5-94)$$

Equation (5-94) can be derived in the same way as equation (5-51), namely by considering the kinetic energy due to the rotation of link i and transforming the angular velocity using the rotation matrix. The inertia tensor $\overline{\mathbf{I}}_i$ is invariant since it depends only on the mass distribution of the link itself. When we evaluate Euler's equation, the inertia tensor $\mathbf{I}_i$ must be obtained for each arm configuration. This requires extra computation time. In the Luh-Walker-Paul's algorithm, all the variables and parameters are expressed in link coordinates so that the additional computation can be eliminated. Namely, instead of representing vectors such as $\mathbf{v}_i$, $\boldsymbol{\omega}_i$, and $\mathbf{a}_i$ with reference to the base coordinate frame, we express them with reference to the coordinate frame fixed to each link, i.e., in link coordinates. To express the equations in link coordinates, we simply replace $\mathbf{v}_i$, $\boldsymbol{\omega}_i$, and the other variables by the ones referred to that link coordinate frame. Further, when a variable referred to frame i is involved in an equation referred to frame $i+1$, it is first premultiplied by the rotation matrix $\mathbf{R}_i^{i+1}$, so that all the variables are expressed with reference to frame $i+1$. In link coordinates, vectors $\mathbf{b}_i$ and $\mathbf{r}_{i,ci}$ are constant, since they are fixed to the link body. Also, if joint i is a revolute joint, the vector $\mathbf{r}_{i,i+1}$ is constant as well.

The computation procedure of the Luh-Walker-Paul's algorithm is summarized in Figure 5-15. The left half of the figure shows the kinematics computation, while the right half shows the dynamics computation. The kinematics computation proceeds downwards, while the dynamics proceeds upwards. The input data of joint motions are transmitted horizontally from the left to the right. The data in the last column are the output joint torques computed through the operations shown by the blocks. The equation numbers in each block indicate the computations to be performed at each stage.

Starting from the top left corner, we first specify the velocity and acceleration of the base link. Note that in this algorithm we can deal with the case when the base frame of the manipulator arm is in motion, if the acceleration of the base frame is known. Note also that the acceleration of gravity is represented as part of the acceleration of the base frame, so that the effect of gravity can be included without extra computation. The first computation block yields the velocity and acceleration of link 1, which are used in the second block for the second step of the computation. Also, the data is transmitted to the right computation block, where

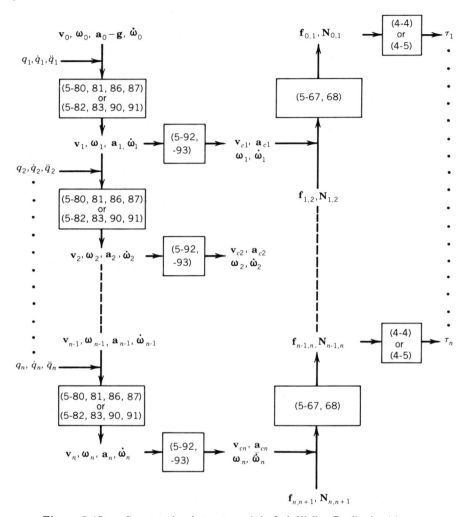

Figure 5-15 :   Computational structure of the Luh–Walker–Paul's algorithm.

the centroidal velocity and acceleration are obtained. The results are further transmitted to the third column, where the Newton-Euler equations are evaluated, and the coupling forces and moments are produced. The result is used to compute the joint torque.

This algorithm is the fastest of existing algorithms for dynamic computation. The number of multiplications required is 852 for a general six degree-of-freedom manipulator arm. It takes 4.5 milliseconds on average to compute the six joint torques on a PDP 11/45 minicomputer using floating point assembly language.

5.4. Research Topics

The derivation of dynamic equations for a manipulator arm is a time-consuming and error-prone process. Automatic generation of the dynamic equations is discussed in (Luh and Lin, 1981; Dillon, 1973; Thomas and Tesar, 1982).

Much effort has been devoted to developing effective procedures to compute the inverse dynamics in real time. A straightforward method is to pre-compute the dynamic equations and use a table look-up technique (Raibert, 1977; Raibert and Horn, 1978). However, this method requires a very large memory size, and is difficult to modify when mass properties change. (Bejczy and Paul, 1981; Bejczy and Lee, 1983) examined for specific robots the relative importance of each dynamic term. (Stephanenko and Vukobratovic, 1976; Orin, *et al.*; Luh, Walker and Paul, 1980-a) devised the recursive Newton-Euler dynamics computation, discussed in Section 5.3, while (Hollerbach, 1980) developed independently the recursive Lagrangian dynamics computation. Later, (Silver, 1982) showed the equivalence between the two approaches. (Hollerbach, 1983) and (Kanade, *et al.*, 1984) further improved the computation efficiency by customizing the dynamic computations to particular robot structures. The recursive computation algorithms were extended to closed-loop kinematic chains by (Luh and Zheng, 1985). (Walker and Orin, 1982) applied the recursive algorithms to dynamics simulation and the *explicit* computation of the inertia matrix and nonlinear torques. The dual-number quaternion algebra, mentioned in Section 2.4, was also used to compute the manipulator dynamics explicitly (Luh and Gu, 1984).

Dynamic analysis has recently been applied to arm linkage design. The goal is to optimize mass properties of link members, as well as their kinematic structure, so that desirable dynamic performance can be achieved. (Asada, 1983) developed the generalized inertia ellipsoid concept, an efficient tool for dynamic analysis and arm design, and applied it to an optimal mass redistribution problem where the arm links are modified to possess isotropic dynamics. (Yoshikawa, 1985; Khatib and Burdick, 1985) extended the dynamic performance evaluation. (Asada and Youcef-Toumi, 1984; Youcef-Toumi and Asada, 1985) studied arm linkage designs to obtain decoupled and configuration-invariant inertia tensors, leading to linear time-invariant arm dynamics, which are easy to control.

Chapter 6
TRAJECTORY CONTROL

6.1. Introduction

This chapter examines how to exploit the preceding developments in the design of effective position or trajectory controllers for robot manipulators.

As seen in Chapter 5, manipulator dynamics can be written in joint-space as

$$\mathbf{H}\ddot{\mathbf{q}} + \mathbf{h} = \boldsymbol{\tau} \tag{6-1}$$

where $\mathbf{q}$ is the vector of joint displacements, $\boldsymbol{\tau}$ is the vector of joint torques, $\mathbf{H} = \mathbf{H}(\mathbf{q})$ is the manipulator inertia matrix, and the nonlinear term $\mathbf{h} = \mathbf{h}(\mathbf{q},\dot{\mathbf{q}};t)$ contains centrifugal, Coriolis, and gravitational forces. The computation of the dynamics requires in general the use of sophisticated on-line computational schemes, as detailed in Section 5.3. The problem is somewhat mitigated in the case of geared manipulators, since the effects of time-varying inertia are reduced by a factor r_j^2, where r_j is the gear ratio, while the interaction torques and nonlinear torques are similarly scaled down by r_j when reflected to the motors. Thus, each line of equation (6-1) is actually of the form

$$\left(h_{Rj} + \frac{h_{aj}}{r_j^2} \right) \ddot{q}_j + \frac{\tau_{cj} + \tau_{nj}}{r_j} = \tau_j \qquad j=1,\ldots,n$$

where h_{Rj} is the (invariant) inertia of the motor rotor including the gearing, h_{aj} is the arm inertia reflected at the joint axis, τ_{cj} is the interactive inertia torque (i.e. the terms in $H_{ij}\ddot{q}_i$, $i \neq j$), and τ_{nj} is the nonlinear torque reflected at the joint axis. In practice, the vector $\mathbf{h}$ of (6-1) also contains friction terms as well as external disturbances; in geared manipulators, the effect of external disturbances is again scaled down by the gear ratio, while the friction terms are generally substantial and thus further contribute to mitigating the nonlinear dynamic effects.   However, the trend in robot design is rapidly shifting to mechanically "cleaner" manipulators.   In direct-drive arms, for instance, high-torque motors are directly coupled to each joint.   This leads to high mechanical stiffness, no backlash, and low friction, but also implies higher sensitivity to external disturbances, as well as full coupling and nonlinear dynamic effects, further enhanced by the very high speeds that the arms are typically capable of reaching.

In order to exploit fully the mechanical capabilities of high-performance robots, it is thus critical to understand how to effectively control dynamics of the form (6-1), and in particular how to properly account for nonlinearities.   In this chapter, we shall detail three approaches to this problem:

(i)   Individual Joint P.I.D.

(ii)   Computed Torque

(iii)   Robust Controller Design

and compare their relative merits.

The first approach, individual joint P.I.D., is essentially the ostrich technique, in that it totally ignores dynamics (6-1) and attempts to control the manipulator by using *local*, decoupled P.I.D.'s at each joint[1]

$$\tau_j := -k_{jD}\,\dot{\tilde{q}}_j - k_{jP}\,\tilde{q}_j - k_{jI}\int_0^t \tilde{q}_j\, dT \qquad\qquad j = 1,\ \cdots\ , n \qquad\qquad (6\text{–}2)$$

where $\tilde{q}_j := q_j - q_{d_j}$ is the position error, and coefficients k_{jP}, k_{jD}, and k_{jI} are positive and "sufficiently large".   Interestingly enough, the approach "almost" works, in the sense that it accomplishes *position control*: after a while , it will get the manipulator where you want it to go.   The proof of this result, discussed in Section 6.2, is strongly based on the structure of Lagrangian dynamics.

---

[1]The notation := is often used in the context of control, and stands for "is defined as".

Clearly, though, something more sophisticated is required in order to obtain good *trajectory tracking* performance. The idea of the *computed torque* method is to define control torque τ using a structure identical to that of dynamics (6-1):

$$\tau := \mathbf{Hu} + \mathbf{h} \tag{6-3}$$

so that substituting (6-3) in (6-1), the problem ideally reduces to that of controlling the simple system

$$\ddot{\mathbf{q}} = \mathbf{u} \tag{6-4}$$

since the inertia matrix $\mathbf{H}$ is positive definite and therefore invertible. Expression (6-4), in turn, merely represents a set of n decoupled double-integrators, each of which can be controlled independently using a simple P.D.

$$u_j := \ddot{q}_{dj} - k_{jD}\dot{\tilde{q}}_j - k_{jP}\,\tilde{q}_j \qquad\qquad j = 1, \cdots, n$$

or P.I.D.

$$u_j := \ddot{q}_{dj} - k_{jD}\dot{\tilde{q}}_j - k_{jP}\,\tilde{q}_j - k_{jI}\int_o^t \tilde{q}_j\,dT \qquad\qquad j = 1, \cdots, n$$

where $\ddot{q}_{dj}$ is the acceleration of the desired trajectory of joint j. The joint torque vector τ of (6-3) may then be computed from the given $\mathbf{u}$ using the effective recursive algorithms developed in Section 5.3 .

The major limitation of this approach, however, is that only *estimates*, $\hat{\mathbf{H}}$ and $\hat{\mathbf{h}}$, of $\mathbf{H}$ and $\mathbf{h}$ are available in practice. Such *parametric uncertainty* is due to inaccuracies on the manipulator mass properties or the torque constants of the actuators, the lack of a good model of friction, unknown loads, perhaps limited on-line computing power, and so on. Thus we can only apply

$$\tau := \hat{\mathbf{H}}\mathbf{u} + \hat{\mathbf{h}} \tag{6-5}$$

instead of (6-3), so that we get

$$\ddot{\mathbf{q}} = (\mathbf{H}^{-1}\,\hat{\mathbf{H}})\mathbf{u} + \mathbf{H}^{-1}(\hat{\mathbf{h}} - \mathbf{h}) \tag{6-6}$$

instead of (6-4). Expression (6-6) shows that the problem is not as simple as (6-4) looked, and in particular may not be adequately handled by standard linear control techniques. A major issue is *robustness*, i.e. how to minimize performance sensitivity to model uncertainty, including both parametric uncertainty *and* high-frequency unmodeled dynamics (such as structural resonant modes, sampling rates, or neglected time-delays). Further, we shall derive explicit *modeling/performance trade-offs*, which shall clarify issues such as the effect of model simplification on tracking performance, reveal the critical importance of an effective selection of control bandwidth, and provide conditions for the effective matching of the computer environment to the arm mechanical capabilities.

The development of this chapter requires familiarity with the concepts of elementary control theory. A brush-up on these concepts and issues is offered in Appendix A.6.1.

6.2. Position Control: Why Local Schemes Work

In this section, we analyse why *local* schemes, consisting of independent linear controllers at each joint, generally represent adequate *position* control schemes for robot manipulators.

We first show that, in the absence of friction or gravity, position control can be achieved by letting

$$\boldsymbol{\tau} := -\mathbf{K}_P\, \tilde{\mathbf{q}} - K_D\, \dot{\tilde{\mathbf{q}}} \tag{6-7}$$

where $\mathbf{K}_P$ and $\mathbf{K}_D$ are any symmetric positive definite matrices, and $\tilde{\mathbf{q}} := \mathbf{q} - \mathbf{q}_d$, with $\mathbf{q}$ the vector of actual joint displacements, and $\mathbf{q}_d$ the fixed vector of desired joint displacements (so that $\dot{\tilde{\mathbf{q}}} = \dot{\mathbf{q}}$).

PROOF 6.2. The proof strongly exploits the structure of Lagrangian dynamics. Based on the development of Section 5.2.3, one can easily show that, in the absence of gravity or friction, the manipulator dynamics (6-1) can be rewritten as:

$$\mathbf{H}\ddot{\mathbf{q}} + \mathbf{C}\dot{\mathbf{q}} = \boldsymbol{\tau}$$

where $\mathbf{C} = \mathbf{C}(\mathbf{q}, \dot{\mathbf{q}})$ is an $n \times n$ matrix and $(\dot{\mathbf{H}} - 2\mathbf{C})$ is *antisymmetric*. Consider then the Lyapunov function candidate

$$V := \frac{1}{2}\left[\tilde{\mathbf{q}}^T \mathbf{K}_P \tilde{\mathbf{q}} + \dot{\mathbf{q}}^T \mathbf{H} \dot{\mathbf{q}} \right]$$

The function V is composed of the manipulator kinetic energy, $1/2\ \dot{\mathbf{q}}^T \mathbf{H} \dot{\mathbf{q}}$, and of a term $1/2\ \tilde{\mathbf{q}}^T \mathbf{K}_P \tilde{\mathbf{q}}$ that accounts for the d.c. gain matrix $\mathbf{K}_P$. Thus, V can be interpreted as the total energy associated with the closed-loop system. Recall that $\mathbf{K}_P$ and $\mathbf{H}$ are symmetric positive definite, so that $V > 0$ except when $\mathbf{q} \equiv \mathbf{q}_d$. Differentiating V with respect to time, we get

$$\dot{V} = \tilde{\mathbf{q}}^T \mathbf{K}_P \dot{\mathbf{q}} + \dot{\mathbf{q}}^T \mathbf{H} \ddot{\mathbf{q}} + 1/2\ \dot{\mathbf{q}}^T \dot{\mathbf{H}} \dot{\mathbf{q}}$$
$$= \tilde{\mathbf{q}}^T \mathbf{K}_P \dot{\mathbf{q}} + \dot{\mathbf{q}}^T (\boldsymbol{\tau} - \mathbf{C}\dot{\mathbf{q}}) + 1/2\ \dot{\mathbf{q}}^T \dot{\mathbf{H}} \dot{\mathbf{q}}$$

Defining control law $\boldsymbol{\tau}$ according to (6-7) thus yields

$$\dot{V} = \tilde{\mathbf{q}}^T \mathbf{K}_P \dot{\mathbf{q}} - \dot{\mathbf{q}}^T (\mathbf{C} + \mathbf{K}_D) \dot{\mathbf{q}} - \dot{\mathbf{q}}^T \mathbf{K}_P \tilde{\mathbf{q}} + 1/2\ \dot{\mathbf{q}}^T \dot{\mathbf{H}} \dot{\mathbf{q}}$$
$$= -\dot{\mathbf{q}}^T (\mathbf{C} + \mathbf{K}_D) \dot{\mathbf{q}} + 1/2\ \dot{\mathbf{q}}^T \dot{\mathbf{H}} \dot{\mathbf{q}}$$

so that

$$\dot{V} = -\dot{\mathbf{q}}^T \mathbf{K}_D \dot{\mathbf{q}} \ \leq \ 0$$

since $(\dot{\mathbf{H}} - 2\mathbf{C})$ is antisymmetric. Note that V (the Lyapunov function itself) depends on $\mathbf{K}_P$, while $\dot{V}$ (the rate of convergence) depends on $\mathbf{K}_D$, which is quite reminiscent of classical results on the effects of proportional and derivative control for single-input linear systems.

There remains to show that the system cannot get "stuck" at a position such that $\dot{V} = 0$ (i.e., $\dot{\mathbf{q}} = \mathbf{0}$) and $\mathbf{q} \neq \mathbf{q}_d$. This is indeed the case, since when $\dot{\mathbf{q}} = \mathbf{0}$ we get:

$$\ddot{\mathbf{q}} = \mathbf{H}^{-1} (\boldsymbol{\tau} - \mathbf{C}\dot{\mathbf{q}}) = -\mathbf{H}^{-1} \mathbf{K}_\mathbf{P} \tilde{\mathbf{q}}$$

which is non-zero as long as $\mathbf{q} \neq \mathbf{q}_d$, since $\mathbf{H}^{-1}\mathbf{K}_P$ is non-singular. V is thus a valid Lyapunov function of the closed-loop system, and the result follows.   **▲▲▲**

Note that the above result is particularly "robust" to uncertainty on the manipulator mass properties, since it does not require any mass parameter estimates at all. In practice, though, $\mathbf{K}_P$ and $\mathbf{K}_D$ must be upper bounded to account for the presence of high-frequency unmodeled dynamics. A particular case of (6-7) is the *local* control law

$$\tau_j = -k_{jP}\,\tilde{q}_j - k_{jD}\,\dot{\tilde{q}}_j \qquad\qquad j=1,\ldots,n \qquad\qquad (6-8)$$

This stability result is not overly surprising: after all, control law (6-8) is merely imitating the behavior that would result from inserting *passive* mechanical elements (namely, a spring and a damper) at each joint. If we now want to account for the presence of gravity torque $\mathbf{g}(\mathbf{q})$ in the system dynamics

$$\mathbf{H\ddot{q}} + \mathbf{C\dot{q}} + \mathbf{g} = \boldsymbol{\tau}$$

we can let

$$\boldsymbol{\tau} := -\mathbf{K}_P\,\tilde{\mathbf{q}} - \mathbf{K}_D\,\dot{\tilde{\mathbf{q}}} + \mathbf{g}(\mathbf{q}) \qquad\qquad (6-9)$$

instead of (6-7). However, such compensation is by nature imperfect since $\mathbf{g}(\mathbf{q})$ in (6-9) is not exactly known (due to uncertainty on link inertias, positions of center of mass, and so on). The effect of Coulomb friction further adds to the final position error, although one could easily show that Proof 6.2 would hold in the presence of viscous friction alone. These problems can be remedied by adding an *integral* control term to (6-8):

$$\tau_j = -k_{jI}\int_0^t \tilde{q}_j\,dT - k_{jP}\,\tilde{q}_j - k_{jD}\,\dot{\tilde{q}}_j \qquad\qquad j=1,\ldots,n \qquad\qquad (6-10)$$

or to (6-9):

$$\tau_j = -k_{jI}\int_0^t \tilde{q}_j\,dT - k_{jP}\,\tilde{q}_j - k_{jD}\,\dot{\tilde{q}}_j - \hat{g}_j(q) \qquad\qquad j=1,\ldots,n$$

$$(6-11)$$

where $\hat{g}_j(\mathbf{q})$ is the available estimate of $g_j(\mathbf{q})$, and k_{jI}, k_{jP} and k_{jD} are within appropriate bounds. The proof of this result is an extension of Proof 6.2 (see Arimoto and Miyazaki, 1983).

Intuitively, the effect of the integral control term in (6-10) or (6-11) is to keep adjusting the control law as long as the (bounded) effects of gravity and Coulomb friction are not exactly compensated.

6.3. Trajectory Control

While local P.I.D.'s are adequate in most position control applications (such as pick-and-place or spot-welding), they generally *overshoot*, i.e. go beyond the specified position before actually stabilizing at it — this is especially true for direct-drive arms where there is very little friction and no gearing to reduce the dynamic effects. Overshooting may be quite undesirable in some cases, for instance if the position-controlled motion is only a prelude to a *compliant* task (i.e. a task that involves contact with the manipulator environment, as detailed in Chapter 7). Employing local schemes then requires to move slowly through a number of intermediate set points, thus considerably delaying the completion of the task. More generally, demanding tasks such as plasma-welding, laser-cutting, or high-speed operations in the presence of obstacles, require effective *trajectory tracking* capabilities.

In order to address such trajectory tracking problems, controller design can become more sophisticated than local P.I.D.'s and *explicitly* account for the nonlinear system dynamics, as seen in the introduction. Namely, given dynamics of the form (6-1)

$$\mathbf{H}\,\ddot{\mathbf{q}} + \mathbf{h} = \boldsymbol{\tau} \tag{6-12}$$

where $\mathbf{q}$ is the vector of joint displacements, a "computed" law of the form (6-5)

$$\boldsymbol{\tau} = \hat{\mathbf{H}}\mathbf{u} + \hat{\mathbf{h}} \tag{6-13}$$

leads to closed-loop dynamics:

$$\ddot{\mathbf{q}} = (\mathbf{H}^{-1}\,\hat{\mathbf{H}})\mathbf{u} + \mathbf{H}^{-1}(\hat{\mathbf{h}} - \mathbf{h}) \tag{6-14}$$

The control challenge is then

(i) To design a (possibly nonlinear) $\mathbf{u}$ in (6-13) to effectively account for

* parameter uncertainty: imprecision on the manipulator mass properties, unknown loads, uncertainty on the load position in the end-effector, inaccuracies on the torque constants of the actuators, friction, and so on.

* the presence of high-frequency *unmodeled* dynamics, such as structural resonant modes, neglected time-delays (in the actuators, for instance), or finite sampling rate; while the frequency content of the control action, i.e., the control *bandwidth*, must be as large as possible so as to most efficiently reduce the effects of parametric uncertainty, it must still be kept low enough not to excite the unmodelled dynamics.

(ii) To explicitly quantify the resulting *modeling/performance trade-offs*. For instance, *local* feedback schemes correspond to $\hat{\mathbf{H}}$ diagonal and $\hat{\mathbf{h}}=\mathbf{0}$. More generally, the effect on tracking performance of discarding any particular term in the dynamic model should be predictable and quantified.

These issues are most easily addressed by using the concepts and notations of *sliding control* theory, which we introduce in this section. The specific design of robust controllers for robot manipulators will be detailed in Section 6.4.

6.3.1. Sliding Surfaces

For notational simplicity, the concepts will first be presented for systems with a single control input. Consider the dynamic system:

$$x^{(n)}(t) = f(\mathbf{X}) + b(\mathbf{X})u(t) + d(t) \tag{6-15}$$

where $u(t)$ is the control input (say the applied torque at a manipulator joint), x is the (scalar) output of interest (say a given joint angle), and $\mathbf{X}=[x,\dot{x}, \cdots ,x^{(n-1)}]^T$ is the state vector. In equation (6-15) the function $f(\mathbf{X})$ (in general nonlinear) is not exactly known, but the *extent of the imprecision* $|\Delta f|$ *on* $f(\mathbf{X})$ *is upper bounded by a known continuous function of* $\mathbf{X}$; similarly control gain $b(\mathbf{X})$ is not exactly known, but is of known sign and is bounded by known, continuous functions of $\mathbf{X}$. Disturbance $d(t)$ is unknown but upper bounded by a known continuous function of $\mathbf{X}$ and t. The control problem is to get the state $\mathbf{X}$ to *track a specific state* $\mathbf{X}_d=[x_d,\dot{x}_d, \cdots ,x_d^{(n-1)}]^T$ *in the presence of model imprecision* on $f(\mathbf{X})$ and $b(\mathbf{X})$, *and of*

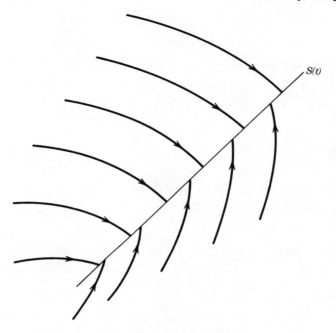

Figure 6-1 : The sliding condition .

disturbances d(t). For this to be achievable using a *finite* control u, we must assume that $\mathbf{X}_d$ is such that

$$\mathbf{X}_d(t=0) \;=\; \mathbf{X}(t=0) \tag{6-16}$$

This assumption is further discussed later in this section and will be relaxed in Section 6.3.2. Let then

$$\tilde{\mathbf{X}} := \mathbf{X} - \mathbf{X}_d = [\tilde{x}, \dot{\tilde{x}}, \cdots , \tilde{x}^{(n-1)}]^T$$

be the tracking error vector.

We define a *time-varying sliding surface S(t)* in the state-space $\mathbf{R}^n$ by the scalar equation $s(\mathbf{X};t) = 0$, where

$$s(\mathbf{X};t) := \left(\frac{d}{dt} + \lambda\right)^{n-1} \tilde{x} \tag{6-17}$$

and λ is a positive constant (design parameter λ will later be interpreted as the desired control bandwidth). Given initial condition (6-16), *the problem of tracking $\mathbf{X} \equiv \mathbf{X}_d$ is equivalent to that* *of remaining on the surface S(t) for all t* > 0 ; indeed $s \equiv 0$ represents a linear differential equation whose unique solution is $\tilde{\mathbf{X}} \equiv 0$, given initial conditions (6-16). Thus the problem of tracking the *n*-dimensional vector $\mathbf{X}_d$ can be reduced to that of keeping the *scalar* quantity *s* at zero. This can in turn be achieved by choosing the control law *u* of (6-15) such that outside of *S(t)*

$$\frac{1}{2}\frac{d}{dt} s^2 (\mathbf{X};t) \leq - \eta|s| \tag{6-18}$$

where η is a positive constant. Inequality (6-18) constrains trajectories to point towards the surface *S(t)*, as illustrated in Figure 6-1, and is referred to as the *sliding condition*.

The idea behind equations (6-17) and (6-18) is to pick-up a well-behaved function of the tracking error, *s*, according to (6-17), and then select the feedback control law *u* in (6-15) such that s^2 remains a Lyapunov function of the closed-loop system, despite the presence of model imprecision and of disturbances. Further, satisfying (6-18) guarantees that if condition (6-16) is not exactly verified, i.e. if $\mathbf{X}|_{t=0}$ is actually off $\mathbf{X}_d|_{t=0}$, the surface *S(t)* will nonetheless be reached in a *finite time* smaller than $|s(t=0)|/\eta$. Indeed, assume for instance that $s(t=0) > 0$, and let t_{reach} be the time required to hit the surface *s*=0. Integrating (6-18) between $t=0$ and $t=t_{\text{reach}}$ leads to

$$0 - s(t=0) = s(t=t_{\text{reach}}) - s(t=0) \leq -\eta(t_{\text{reach}} - 0)$$

which implies that

$$t_{\text{reach}} \leq s(t=0)/\eta$$

One would obtain a similar result starting with $s(t=0) < 0$, and thus

$$t_{\text{reach}} \leq |s(t=0)|/\eta$$

Further, definition (6-17) implies that once on the surface, tracking error tends exponentially to zero, with a time constant $(n - 1)/\lambda$.

The controller design procedure then consists of two steps. First, as will be illustrated in Section 6.3.2, a feedback control law *u* is selected so as to verify sliding condition (6-18).

However, in order to account for the presence of modeling imprecision and of disturbances, such a control law has to be *discontinuous across S(t)*, which leads to control *chattering*. Now chattering is undesirable in practice, since it involves high control activity and further may excite high-frequency dynamics neglected in the course of modeling (such as unmodeled structural modes, neglected time-delays, and so on). Thus, in a second step detailed in Section 6.3.3, the *discontinuous control law u is suitably smoothed to achieve an optimal trade-off between control bandwidth and tracking precision*: while the first step accounts for parametric uncertainty, the second step achieves robustness to high-frequency unmodeled dynamics.

6.3.2. Perfect Tracking Using Switched Control Laws

Given the bounds on uncertainties on $f(\mathbf{X})$ and $b(\mathbf{X})$, and on disturbances $d(t)$, constructing a control law to verify sliding condition (6-18) is straightforward, as we now illustrate on simple examples.

<u>EXAMPLE 6-1</u>

Consider the second-order system

$$\ddot{x} = f + u \qquad\qquad (6-19)$$

where u is the control input, x is the (scalar) output of interest, and dynamics f (possibly nonlinear or time-varying) is not exactly known, but estimated as $\hat{f}$. The estimation error on f is assumed to be *bounded* by some known function $F = F(x, \dot{x})$:

$$|\hat{f} - f| \leq F \qquad\qquad (6-20)$$

In order to have the system track $x(t) \equiv x_d(t)$, we define a sliding surface $s = 0$ according to (6-17), namely:

$$s := \left(\frac{d}{dt} + \lambda\right) \tilde{x} = \dot{\tilde{x}} + \lambda \tilde{x}$$

We then have:

$$\dot{s} = \ddot{x} - \ddot{x}_d + \lambda \dot{\tilde{x}} = f + u - \ddot{x}_d + \lambda \dot{\tilde{x}} \qquad\qquad (6-21)$$

The best approximation $\hat{u}$ of a continuous control law that would achieve $\dot{s} = 0$ is thus

$$\hat{u} = -\hat{f} + \ddot{x}_d - \lambda\dot{\tilde{x}} \qquad (6-22)$$

In order to satisfy sliding condition (6-18) despite uncertainty on the dynamics f, we add to $\hat{u}$ a term *discontinuous* across the surface $s = 0$:

$$u := \hat{u} - k\ sgn(s) \qquad (6-23)$$

where *sgn* is the sign function:

$$sgn(s) = +1 \qquad \text{if } s > 0$$
$$sgn(s) = -1 \qquad \text{if } s < 0$$

By choosing $k = k(x,\dot{x})$ in (6-23) to be large enough, we can now guarantee that (6-18) is verified. Indeed, we have from (6-21)-(6-23)

$$\frac{1}{2}\frac{d}{dt}s^2 = \dot{s}\cdot s = [\,f - \hat{f} - k\ sgn(s)\,]\ s = (f - \hat{f})s - k|s|$$

so that, letting

$$k := F + \eta \qquad (6-24)$$

we get from (6-20)

$$\frac{1}{2}\frac{d}{dt}s^2 \leq -\eta|s|$$

as desired. Note from (6-24) that control discontinuity k across the surface $s = 0$ increases with the extent of parametric uncertainty. Also, remark that $\hat{f}$ and F need not depend only on x or $\dot{x}$, but may more generally be functions of any *measured* variables external to system (6-19).

$$\Delta\Delta\Delta$$

EXAMPLE 6-2

An equivalent result would be obtained by using integral control, i.e. formally letting $(\int_0^t \tilde{x} dT)$ be the variable of interest. The system (6-19) is third-order relative to this variable, and (6-17) gives:

$$s := \left(\frac{d}{dt} + \lambda\right)^2 \left(\int_0^t \tilde{x} dT\right) = \dot{\tilde{x}} + 2\lambda\tilde{x} + \lambda^2 \int_0^t \tilde{x} dT$$

We then obtain, instead of (6-22),

$$\hat{u} = -\hat{f} + \ddot{x}_d - 2\lambda\dot{\tilde{x}} - \lambda^2\tilde{x}$$

with (6-23), (6-24) formally unchanged. Note that $\int_0^t \tilde{x} dT$ can be replaced by $\int^t \tilde{x} dT$, i.e. the integral can be defined to within a constant. The constant can be chosen to obtain $s(t{=}0) = 0$ *regardless of* $\mathbf{X}_d(t{=}0)$, by letting

$$s := \dot{\tilde{x}} + 2\lambda\tilde{x} + \lambda^2 \int_0^t \tilde{x} dT - \dot{\tilde{x}}(0) - 2\lambda\tilde{x}(0) \qquad \blacktriangle\blacktriangle\blacktriangle$$

EXAMPLE 6-3

Assume now that (6-19) is replaced by

$$\ddot{x} = f + bu \qquad\qquad (6\text{-}25)$$

where (possibly nonlinear or time-varying) control gain b is known only to within a certain *gain margin* $\beta{=}\beta(\mathbf{X})$:

$$\beta^{-1} \leq \frac{\hat{b}}{b} \leq \beta \qquad\qquad (6\text{-}26)$$

where $\hat{b}$ is the available estimate of gain b. With s and $\hat{u}$ defined as in Examples 6-1 or 6-2, one can then easily show that the control law

$$u = \hat{b}^{-1}[\hat{u} - k\ sgn(s)] \tag{6-27}$$

with

$$k = \beta(F + \eta) + (\beta-1)\,|\hat{u}| \tag{6-28}$$

satisfies the sliding condition. Indeed, using (6-27) in the expression of $\dot{s}$ leads to

$$\dot{s} = (f - b\hat{b}^{-1}\ \hat{f}) + (1 - b\hat{b}^{-1})(-\ddot{x}_d + \lambda\dot{\tilde{x}}) - b\hat{b}^{-1}k\ sgn(s)$$

so that k must verify

$$k \geq |\hat{b}b^{-1}f - \hat{f} + (\hat{b}b^{-1} - 1)(-\ddot{x}_d + \lambda\dot{\tilde{x}})| + \eta\hat{b}b^{-1}$$

Since $f = \hat{f} + (f - \hat{f})$, where $|f - \hat{f}| \leq F$, this in turn leads to

$$k \geq \hat{b}b^{-1}\,F + \eta b\hat{b}^{-1} + |\hat{b}b^{-1} - 1|\cdot|\hat{f} - \ddot{x}_d + \lambda\dot{\tilde{x}}|$$

and thus to (6-28). Note that control discontinuity has been increased in order to account for uncertainty on control gain b.   ◮◮◮

EXAMPLE 6-4

Still for the system (6-25), assume that we have

$$\beta_{min} \leq \frac{\hat{b}}{b} \leq \beta_{max}$$

instead of (6-26). Uncertainty on control gain b can be put back in the form (6-26) by letting

$$\beta := (\beta_{max}/\beta_{min})^{1/2}$$

and replacing estimated gain $\hat{b}$ by $(\beta_{min}\beta_{max})^{-1/2}\ \hat{b}$. Similarly, knowing that

$$b_{min} \leq b \leq b_{max}$$

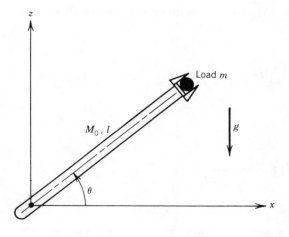

Figure 6-2 : One-link manipulator.

leads to

$$\hat{b} := (b_{min}b_{max})^{1/2} \quad ; \quad \beta := (b_{max}/b_{min})^{1/2} \qquad \Delta\Delta\Delta$$

Example 6-5

Consider the simple one-link manipulator of Figure 6-2. The manipulator is required to move unknown loads along desired trajectories $\theta(t) \equiv \theta_d(t)$. With M_o the manipulator mass, l its length, and m the unknown mass of the load, the system dynamics can be written

$$I \ddot{\theta} + f_D\dot{\theta} + \mu g \, cos(\theta) = \tau$$

where

$$I = \frac{1}{3}M_o l^2 + ml^2$$

$$M = M_o + m$$

$$\mu = l\frac{M_o}{2} + lm$$

and there is also a 50% uncertainty on the damping term:

$$f_D = \hat{f}_D \pm 50\% \; \hat{f}_D$$

Assume further that we know that $0 \le m \le M_o$. From Example 6-4, the resulting gain margin is

$$\beta = \left((4/3 \ M_o l^2)/(1/3 \ M_o l^2) \right)^{1/2} = \quad 2$$

and the corresponding gain estimate is

$$\hat{b} = \left((4/3 \ M_o l^2)(1/3 \ M_o l^2) \right)^{-1/2} = \frac{3}{2 \ M_o l^2}$$

Using for simplicity $M_o/2$ as our estimate of the load mass $\hat{m}$, so as to minimize the uncertainty on μ (actually, one could still reduce the necessary control discontinuity by explicitly accounting for the dependency between I and μ), the control law is then

$$\tau = \hat{b}^{-1} [\hat{u} - k \ sgn(s)]$$

where

$$s = \dot{\tilde{\theta}} + \lambda \tilde{\theta} \tag{6--29}$$

$$\hat{u} = \hat{b} \left(\hat{f}_D \dot{\theta} + l M_o \ g \ cos(\theta) \right) + \ddot{\theta}_d - \lambda \dot{\tilde{\theta}} \tag{6--30}$$

and

$$k = \hat{b} \left(\hat{f}_D |\dot{\theta}|/2 \ + \ \frac{l M_o}{2} \ g |cos(\theta)| \ \right) + \beta \eta + (\beta - 1) \ |\ddot{\theta}_d - \lambda \dot{\tilde{\theta}}|$$

An equivalent result is obtained by letting

$$s := \dot{\tilde{\theta}} + 2\lambda \tilde{\theta} + \lambda^2 \int^t \tilde{\theta} \, dT$$

instead of (6-29), and thus replacing (6-30) by

$$\hat{u} = \hat{b}\left(\hat{f}_D\dot{\theta} + lM_o \; g \; cos(\theta)\right) + \ddot{\theta}_d - 2\lambda\dot{\tilde{\theta}} - \lambda^2\tilde{\theta}$$

Control discontinuity gain k is modified accordingly.   △△△

6.3.3. Continuous Control Laws to Approximate Switched Control

As remarked in the preceding examples, control laws which satisfy sliding condition (6-18), and thus lead to "perfect" tracking in the face of model uncertainty, are discontinuous across the surface $S(t)$, thus leading to control *chattering*. Chattering is generally highly undesirable in practice, since it involves extremely high control activity, and further may excite high-frequency dynamics neglected in the course of modeling. We can remedy this situation by smoothing out the control discontinuity in a thin *boundary layer* neighboring the switching surface:

$$B(t) = \{\mathbf{X}, |s(\mathbf{X};t)| \leq \Phi\} \quad ; \quad \Phi > 0$$

where Φ is the boundary layer *thickness*, and $\epsilon := \Phi/\lambda^{n-1}$ is the boundary layer *width* (Figure 6-3.a). This is achieved by choosing *outside* of $B(t)$ control law u as before (i.e., satisfying sliding condition (6-18)), which guarantees boundary layer attractiveness hence positive invariance: all trajectories starting inside $B(t{=}0)$ remain inside $B(t)$ for all $t \geq 0$; and then interpolating u inside $B(t)$ — for instance, replacing in the expression of u the term $sgn(s)$ by s/Φ, inside $B(t)$, as illustrated in Figure 6-3.b.

We first show that this operation leads to *tracking to within a guaranteed precision* ϵ, and more generally guarantees that for all trajectories starting inside $B(t{=}0)$

$$|\tilde{x}^{(i)}(t)| \leq (2\lambda)^i \epsilon \qquad i = 0, \dots, n-1 \tag{6-31}$$

Indeed, assume first that $\tilde{\mathbf{X}}(0) = \mathbf{0}$. By definition (6-17), tracking error $\tilde{x}$ is obtained from s through a sequence of first-order lowpass filters (Figure 6-4.a, where $p = (d/dt)$ is the Laplace operator). Let then y_1 be the output of the first filter. We have

$$y_1(t) = \int_0^t e^{-\lambda(t-T)}s(T) \; dT$$

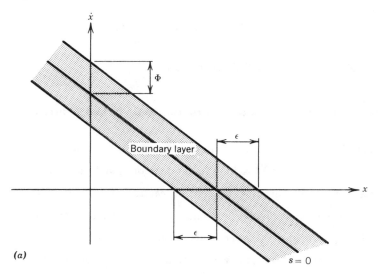

(a)

Figure 6-3a : Construction of the boundary layer in the case $n = 2$.

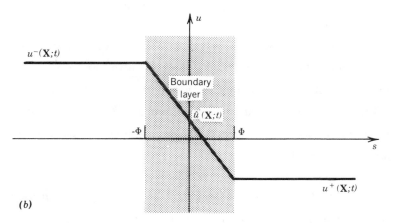

(b)

Figure 6-3b : Control law interpolation in the boundary layer.

From $|s| \leq \Phi$ we thus get

$$|y_1(t)| \leq \Phi \int_0^t e^{-\lambda(t-T)}dT = (\Phi/\lambda)(1-e^{-\lambda t}) \leq \Phi/\lambda$$

We can apply the same reasoning to the second filter, and so on, all the way to $y_{n-1} = \tilde{x}$. We then get

$$|\tilde{x}| \leq \Phi/\lambda^{n-1} = \epsilon$$

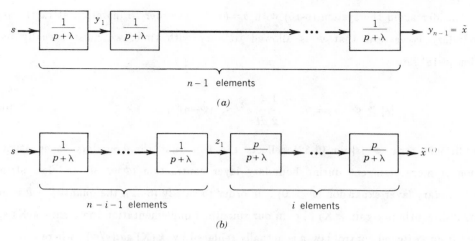

$n-1$ elements

(a)

$n-i-1$ elements i elements

(b)

Figure 6-4 : Derivation of bounds (6-31).

Similarly, $\tilde{x}^{(i)}$ can be thought of as obtained through the sequence of Figure 6-4.b. From the previous result, one has $|z_1| \leq \Phi/\lambda^{n-1-i}$. Further, noting that

$$\frac{p}{p+\lambda} = \frac{p+\lambda-\lambda}{p+\lambda} = 1 - \frac{\lambda}{p+\lambda}$$

one sees that the sequence of Figure 6-4.b implies

$$|\tilde{x}^{(i)}| \leq \left(\frac{\Phi}{\lambda^{n-1-i}}\right)\left(1 + \frac{\lambda}{\lambda}\right)^i = (2\lambda)^i\epsilon$$

i.e. bounds (6-31). Finally, in the case that $\tilde{\mathbf{X}}(0) \neq \mathbf{0}$, bounds (6-31) are obtained asymptotically, i.e. within a short time-constant $(n\text{-}1)/\lambda$.

Further, as we now show, *the smoothing of control discontinuity inside B(t) essentially assigns a lowpass filter structure to the local dynamics of the variable s*, thus eliminating chattering. Recognizing this filter structure then allows us to tune up the control law so as to achieve a trade-off between tracking precision and robustness to unmodeled dynamics. Boundary layer thickness Φ is made to be *time-varying* and is monitored so as to always exploit the maximum control bandwidth available. The development is first detailed for the case $\beta=1$ (no gain margin), and then generalized.

Consider again the system (6-15) with $b = \hat{b} = 1$. In order to maintain attractiveness of the boundary layer now that Φ is allowed to vary with time, we must actually modify condition (6-18) into

$$|s| \geq \Phi \quad \Rightarrow \quad \frac{1}{2}\frac{d}{dt}s^2 \leq (\dot{\Phi} - \eta)|s| \qquad (6-32)$$

The additional term $\dot{\Phi}\,|s|$ in (6-32) reflects the fact that the boundary layer attraction condition is more stringent during boundary layer contraction ($\dot{\Phi} < 0$) and less stringent during boundary layer expansion ($\dot{\Phi} > 0$). In order to satisfy (6-32), the quantity $-\dot{\Phi}$ is added to control discontinuity gain $k(\mathbf{X})$, i.e. in our smoothed implementation the term $-k(\mathbf{X})\,sgn(s)$ obtained from switched control law u is actually replaced by $\overline{k}(\mathbf{X})\,sat(s/\Phi)$, where

$$\overline{k}(\mathbf{X}) := k(\mathbf{X}) - \dot{\Phi} \qquad (6-33)$$

and sat is the saturation function:

$$sat(y) = y \qquad \text{if } |y| \leq 1$$
$$sat(y) = sgn(y) \qquad \text{otherwise}$$

Accordingly, control law u becomes:

$$u = \hat{u} - \overline{k}(\mathbf{X})\,sat(s/\Phi)$$

Let us now consider the system trajectories *inside the boundary layer*, where they lie by construction: they can be expressed directly in terms of the variable s as

$$\dot{s} = -\overline{k}(\mathbf{X})\,s/\Phi + \left(-\Delta f(\mathbf{X}) + d(t)\right) \qquad (6-34)$$

where $\Delta f := \hat{f} - f$. Now since $\overline{k}$ and Δf are continuous in $\mathbf{X}$, we can exploit (6-31) to rewrite (6-34) in the form

$$\dot{s} = -\overline{k}(\mathbf{X}_d)\,s/\Phi + \left(-\Delta f(\mathbf{X}_d) + d(t) + O(\epsilon)\right) \qquad (6-35)$$

We see from (6-35) that *the variable s* (which is a measure of the algebraic distance to the

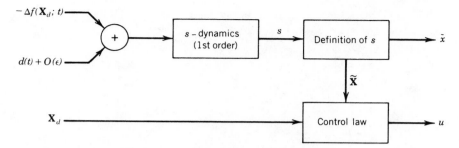

Figure 6-5 : Dynamic structure of the closed-loop system.

surface $S(t)$ *is the output of a stable first-order filter,* whose dynamics only depends on the desired state $\mathbf{X}_d(t)$, and whose inputs are "perturbations", to the first order, namely disturbance $d(t)$ and uncertainty $\Delta f(\mathbf{X}_d)$. Chattering is thus indeed eliminated, as long as high-frequency unmodeled dynamics are not excited. Conceptually, the dynamic structure of the closed-loop system can be summarized by Figure 6-5: perturbations are filtered according to (6-35) to give s, which in turn provides tracking error $\tilde{x}$ by further lowpass filtering, according to definition (6-17). Control action u is a function of $\mathbf{X}$ and $\mathbf{X}_d$. Now, since λ is the break-frequency of filter (6-17), it must be chosen to be "small" with respect to high-frequency unmodeled dynamics (such as unmodeled structural modes or neglected time delays). Further, we can now tune boundary layer thickness Φ so that (6-35) also represents a first-order filter of bandwidth λ. It suffices to let

$$\frac{\overline{k}(\mathbf{X}_d)}{\Phi} := \lambda \tag{6-36}$$

which can be written from (6-33) as

$$\dot{\Phi} + \lambda\Phi = k(\mathbf{X}_d) \tag{6-37}$$

Equation (6-37) defines the desired time-history of boundary layer thickness Φ, and, in the light of Figure 6-5, shall be referred to as the *balance condition.* Intuitively, it amounts to tune up the closed-loop system so that it mimics an n-th order critically damped system. Further, definition (6-33) can then be rewritten as

$$\overline{k}(\mathbf{X}) := k(\mathbf{X}) - k(\mathbf{X}_d) + \lambda\Phi \tag{6-38}$$

In the case that $\beta \neq 1$, one can easily show that (6-37) and (6-38) become (with $\beta_d = \beta(\mathbf{X}_d)$)

$$k(\mathbf{X}_d) \geq \frac{\lambda\Phi}{\beta_d} \quad => \quad \dot{\Phi} + \lambda\Phi = \beta_d k(\mathbf{X}_d) \tag{6-39}$$

$$k(\mathbf{X}_d) \leq \frac{\lambda\Phi}{\beta_d} \quad => \quad \dot{\Phi} + \frac{\lambda\Phi}{\beta_d^2} = k(\mathbf{X}_d)/\beta_d \tag{6-40}$$

and

$$\overline{k}(\mathbf{X}) := k(\mathbf{X}) - k(\mathbf{X}_d) + \lambda\Phi/\beta_d \tag{6-41}$$

with initial condition $\Phi(0)$ defined as:

$$\Phi(0) := \beta_d \, k(\mathbf{X}_d(0);0)/\lambda \tag{6-42}$$

Indeed, in order to satisfy (6-32) in the presence of uncertainty β on the control gain we let

$$\dot{\Phi} > 0 \quad => \quad \overline{k}(\mathbf{X}) = k(\mathbf{X}) - \dot{\Phi}/\beta \tag{6-43}$$

$$\dot{\Phi} < 0 \quad => \quad \overline{k}(\mathbf{X}) = k(\mathbf{X}) - \beta\,\dot{\Phi} \tag{6-44}$$

Further, the balance condition can be written, instead of (6-36), as

$$\left(\frac{\overline{k}(\mathbf{X}_d)}{\Phi}\right)\left(\frac{b(\mathbf{X}_d)}{\hat{b}(\mathbf{X}_d)}\right)_{\max} = \lambda \tag{6-45}$$

that is,

$$\overline{k}(\mathbf{X}_d) = \lambda\Phi/\beta_d$$

Applying this relation to (6-43), (6-44) leads to the desired behavior of $\dot{\Phi}$:

$$\dot{\Phi} > 0 \quad => \quad \frac{\lambda\Phi}{\beta_d} = k(\mathbf{X}_d) - \dot{\Phi}/\beta_d \tag{6-46}$$

$$\dot{\Phi} < 0 \quad => \quad \frac{\lambda\Phi}{\beta_d} = k(\mathbf{X}_d) - \beta_d\dot{\Phi} \tag{6-47}$$

that is

$$\dot{\Phi} > 0 \quad => \quad \dot{\Phi} = (\beta_d k(\mathbf{X}_d) - \lambda\Phi)$$

$$\dot{\Phi} < 0 \quad => \quad \dot{\Phi} = \frac{1}{\beta_d^2}(\beta_d k(\mathbf{X}_d) - \lambda\Phi)$$

which we can rewrite as (6-39) and (6-40). Finally (6-41) is obtained by remarking that

$$\overline{k}(\mathbf{X}) = \left(\overline{k}(\mathbf{X}) - \overline{k}(\mathbf{X}_d)\right) + \overline{k}(\mathbf{X}_d) = \left(k(\mathbf{X}) - k(\mathbf{X}_d)\right) + \lambda\Phi/\beta_d$$

Note that the balance conditions (6-39) and (6-40) imply that Φ and thus $\tilde{\mathbf{X}}$ are bounded for bounded $\mathbf{X}_d$.

The balance conditions have a simple and intuitive physical interpretation: neglecting time constants of order $1/\lambda$, they imply that

$$\lambda^n \epsilon \approx \beta_d\, k(\mathbf{X}_d) \tag{6-48}$$

that is

(bandwidth)^n × (tracking precision)

$\approx$ (parametric uncertainty measured along the desired trajectory)

Such trade-off is quite similar to the situation encountered in linear time-invariant systems, but here applies to the more general nonlinear system structure (6-15), all along the desired trajectory. In particular, it shows that *the balance conditions specify the best tracking performance attainable, given the desired control bandwidth and the extent of parameter uncertainty.* Similarly, expression (6-28) allows one to easily quantify the trade-off between accuracy and speed along a given path, for instance. The important implications of (6-48) for robot controller design shall be discussed in Section 6.5.

REMARKS:

(i) The desired trajectory $\mathbf{X}_d$ must itself be chosen smooth enough not to excite the high-frequency unmodeled dynamics.

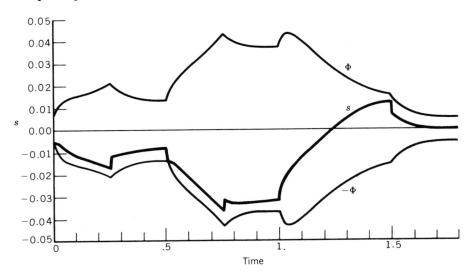

Figure 6-6 : Example of *s* - trajectory.

(ii) An argument similar to that of the above discussion shows that the choice of dynamics (6-17) used to define sliding surfaces is the "best-conditioned" among linear dynamics, in the sense that it guarantees the best tracking performance given the desired control bandwidth and the extent of parameter uncertainty.

(iii) The *s-trajectory*, i.e. the variation of s/Φ with time, is a compact descriptor of the closed-loop behavior: control activity directly depends on s/Φ, while by definition (6-17) tracking error $\tilde{x}$ is merely a filtered version of *s*. Further, the *s-trajectory represents a time-varying measure of the validity of the assumptions on model uncertainty*. Similarly, boundary layer thickness Φ describes the evolution of dynamic model uncertainty with time. It is thus particularly informative to plot $s(t)$, $\Phi(t)$, and $-\Phi(t)$ on a single diagram, as illustrated in Figure 6-6.

(iv) In certain applications, control input u is an electrical signal rather than a mechanical force or acceleration. Control chattering may then be acceptable provided it is *beyond* the frequency range of the relevant unmodeled dynamics. Electric motors, for instance, are generally controlled using pulse-width modulation (PWM) at very high switching rates. For such applications, and provided that the necessary computations (including both control law and state estimation) can be handled on-line at a high enough rate, or implemented using analog circuitry, pure sliding mode control using switched control laws may be a viable and extremely high-performance alternative. The practical aspects linked to such implementation represent an interesting area of current research (Bondarev *et al.*, 1985).

We now describe the application of the above development to the robust feedback control of robot manipulators.

6.4. Robust Trajectory Control For Robot Manipulators

Much effort has been devoted to developing efficient procedures for real-time computation of manipulator dynamics. As seen in Section 5.3 , a substantial improvement in computation efficiency is obtained by using *recursive* algorithms to generate the torques required to support a desired motion (inverse dynamics), achieving linear variation of the computation complexity with the number of links. However, while the inverse dynamics computes the torques theoretically needed to compensate for nonlinearities and follow a specific trajectory, assuming an *exact* model, the joint accelerations are affected not only by control torques and known dynamics, but also by *disturbances* (such as Coulomb and viscous friction) and *modeling errors* : parametric uncertainties such as inaccuracies on inertias, geometry, torque constants of the actuators, mass of the load and its exact position in the end-effector; and high-frequency unmodeled dynamics, such as unmodeled structural modes or neglected time-delays. Now there are several reasons for insisting on *explicit* robustness guarantees for the robot control system. The first is obvious: control instabilities are unpleasant, especially at high speeds. Conversely, *guaranteed robustness properties allow one to design simple controllers*. Consider for instance a 6 d.o.f. manipulator, composed of a 3 d.o.f. arm and a 3 d.o.f. hand. As seen in Section 2.3.3, a kinematic decoupling between hand and arm can be achieved by having the three rotational axis of the wrist intersect at a point (spherical wrist). Now physically it seems natural to seek a similar decoupling at the dynamic level, in other words to be able to consider motions of arm and hand as "disturbances" to one another (each of these disturbances being possibly further decomposed into an average, quickly estimated term to be directly compensated for, and a genuine perturbation term to be accounted for by the controller robustness). Further, desired bandwidth is likely to be much larger for the hand than for the arm itself. Robust controller design does allow such natural reduction of the original problem into two lower-order control problems: in the case of a 6 d.o.f. manipulator, both of these problems are amenable to closed-form, pencil and paper treatment, thus allowing to maintain clear physical insight and exploit engineering judgment all along the design and implementation process.

Depending on the structure of model uncertainty and on the type of application considered, the control techniques of Section 6.3 may be applied in various ways, as shall be further detailed in Chapter 7. In this section we assume that the desired trajectories are specified in joint space. Also, as discussed above, if N is the total number of d.o.f. of a wrist-partitioned manipulator and N_H is the number of d.o.f. of the wrist and hand, the first $N-N_H$ and the last N_H degress of freedom of the robot may be treated separately. Thus, in the sequel, system size n may refer to $N-N_H$, N_H, or N.

6.4.1. Controller Structure

Let λ be the desired control bandwidth, chosen small enough not to excite unmodeled structural modes or interfere with neglected time-delays (as we shall further detail in Section 6.5), and let

$$\hat{\mathbf{H}}(\mathbf{q})\ddot{\mathbf{q}} + \hat{\mathbf{h}} = \boldsymbol{\tau} \tag{6-49}$$

be the available *model* of manipulator (6-1). Discrepancies between (6-1) and (6-49) may arise from several factors: imprecisions on the manipulator geometry or inertias, uncertainties on the friction terms or the loads, inaccuracies on the torque constants of the actuators, on-line computation limitations, or purposeful model simplification. The sliding controller for manipulator (6-1) takes the form:

$$\boldsymbol{\tau} = \hat{\mathbf{H}}\mathbf{u} + \hat{\mathbf{h}} \tag{6-50}$$

where the components u_i $(i = 1, \ldots, n)$ of vector $\mathbf{u}$ are defined as:

$$u_i = G_i(\mathbf{q})\left[\hat{u}_i - \overline{k}_i(\mathbf{q},\dot{\mathbf{q}}) \ sat(s_i/\varPhi_i)\right] \tag{6-51}$$

In equation (6-51), $\overline{k}_i(\mathbf{q},\dot{\mathbf{q}})$ and Φ_i are defined according to the dynamic balance conditions (6-39)-(6-42), and

$$\hat{u}_i := \ddot{q}_{di} - 2\lambda\dot{\tilde{q}}_i - \lambda^2\tilde{q}_i$$

where $\ddot{q}_{di}$ is the acceleration of the desired trajectory at joint i, and $\tilde{q}_i$ is the tracking error at joint i. The surfaces s_i in (6-51) are set to be

$$s_i := \dot{\tilde{q}}_i + 2\lambda\tilde{q}_i + \lambda^2\int^t \tilde{q}_i(T)dT \qquad (6-52)$$

so as to use integral control and reduce the effects of friction. The term $\int^t \tilde{q}_i(T)dT$ is defined such that $s_i(t{=}0) = 0$, as in Example 6-2 of Section 6.3.2. The scaling factor $G_i(\mathbf{q})$ of (6-51) and the bounds on model uncertainty used in the computation of $\overline{k}_i(\mathbf{q},\dot{\mathbf{q}})$ are discussed in the next section. Note that the major difference with a mere computed torque, which would involve a linear controller based on the approximation (6-4), is the presence of the robustifying terms in $G_i(\mathbf{q})\ \overline{k}_i(\mathbf{q},\dot{\mathbf{q}})\ sat(s_i/\Phi_i)$ in expression (6-52), which allow us to maintain stability and optimize performance in the face of model uncertainty.

6.4.2. Practical Evaluation of Parametric Uncertainties

Let us now evaluate explicit bounds on parametric uncertainty, so as to generate a set of control discontinuity gains $k_i(\mathbf{q},\dot{\mathbf{q}})$ as in section 6.3.2; these gains will then be fed into the dynamic balance conditions in order to compute modified gains $\overline{k}_i(\mathbf{q},\dot{\mathbf{q}})$ and boundary layer thicknesses Φ_i of equation (6-51). A simplified, easily implementable approach to such evaluation is as follows: define two sets of n-vectors, $\mathbf{L}_j\ (j{=}1,\ldots,n)$ and $\Delta\mathbf{H}_j\ (j{=}1,\ldots,n)$ by

$$\mathbf{H}^{-1} =: [\mathbf{L}_1 \ldots \mathbf{L}_n]$$

$$\hat{\mathbf{H}} - \mathbf{H} =: [\Delta\mathbf{H}_1 \ldots \Delta\mathbf{H}_n]$$

The vectors $\mathbf{L}_j$ and $\Delta\mathbf{H}_j$ are functions of configuration $\mathbf{q}$ only. Substituting control law (6-50) into robot dynamics (6-49) the dynamics of the closed loop system can be written:

$$\ddot{\mathbf{q}} = (\mathbf{I} + \mathbf{H}^{-1}\Delta\mathbf{H})\mathbf{u} + \mathbf{H}^{-1}\Delta\mathbf{h}$$

where $\mathbf{I}$ is the $n \times n$ identity matrix and $\Delta\mathbf{h} = \hat{\mathbf{h}} - \mathbf{h}$ reflects the effects of unmodeled gravitational loads, friction, or model simplification. Assume now that

$$\mathbf{L}_i^T \cdot \hat{\mathbf{H}}_i > 0 \qquad i = 1, \ldots, n \qquad (6-53)$$

This condition is quite mild: recalling that $\mathbf{H}^{-1}$ is symmetric, it simply means that u_i contributes to $\ddot{q}_i$ with a predictable sign — actually, inequality (6-53) can always be satisfied by letting $\hat{\mathbf{H}}$ be a (possibly time-varying) positive definite *diagonal* matrix. In other words, if because of parameter uncertainty the computed $\hat{\mathbf{H}}$ is not even able to satisfy (6-53), one should rather use a simple diagonal matrix (composed of the diagonal elements of the original $\hat{\mathbf{H}}$, for instance). Define then a set of scalars $\beta_i^{\min} = \beta_i^{\min}(\mathbf{q})$, $\beta_i^{\max} = \beta_i^{\max}(\mathbf{q})$ such that

$$0 \leq \beta_i^{min} \leq \mathbf{L}_i^T \cdot \hat{\mathbf{H}}_i \leq \beta_i^{max} \qquad i = 1, \ldots, n \qquad (6-54)$$

The multiplier $G_i = G_i(\mathbf{q})$ in (6-50) then simply centers the estimated gain, as in Example 6-4 of section 6.3.2:

$$G_i := (\beta_i^{min}\,\beta_i^{max})^{-1/2} \qquad (6-55)$$

The corresponding gain margin is

$$\beta_i := (\beta_i^{max}/\beta_i^{min})^{1/2} \qquad (6-56)$$

The control discontinuity gains $k_i(\mathbf{q},\dot{\mathbf{q}})$ must then verify the simplified conditions (for $i = 1, \cdots, n$)

$$k_i(\mathbf{q},\dot{\mathbf{q}}) \geq \beta_i\left[(1-\beta_i^{-1})|\hat{u}_i| + (\mathbf{L}_i^+)^T\Delta\mathbf{h}^+ + \sum_{j \neq i} G_j\,|\ddot{q}_{dj}|\cdot|\mathbf{L}_i^T\cdot\hat{\mathbf{H}}_j| + \eta_i\right] \qquad (6-57)$$

where the η_i are the positive constants used in the expression (6-18) of each sliding condition and, by definition, the components of $\Delta \mathbf{h}^+$ and $\mathbf{L}_i^+$ are the absolute values of the components of $\Delta \mathbf{h}$.

Conditions (6-54) and (6-57) are fairly straightforward to satisfy for $n = 3$ (hence in particular for a 6 d.o.f. wrist-partitioned robot) since closed form dynamics remain manageable analytically — especially, an off-line worst-case analysis may be sufficient in practice. In the general case, an adequate first-order approximation is to evaluate bounds (6-54) and (6-57) by using estimates $\hat{\mathbf{L}}_i^+$ (obtained by inverting $\hat{\mathbf{H}}$) in place of the actual $\mathbf{L}_i^+$. Similarly, approximate upper bounds on the components of $\Delta \mathbf{H}_i^+$ or $\Delta \mathbf{h}^+$ may be easily expressed in terms of $\hat{\mathbf{H}}_i^+$ or $\hat{\mathbf{h}}^+$ (with obvious notations) — for instance, if the only source of uncertainty on estimated inertia matrix $\hat{\mathbf{H}}$ is that manipulator inertias are known to within a 10 percent precision, one may use in practice the approximation $\Delta \hat{\mathbf{H}}_i^+ \leq 10\% \; \mathbf{H}_i^+$, where the inequality is understood componentwise. Note that bounds (6-54) and (6-57) can be computed systematically based on parametric uncertainty structure.

REMARKS:

(i) Complete expressions of the $k_i(\mathbf{q}, \dot{\mathbf{q}})$ are derived in Appendix A.6.2.

(ii) A well-designed controller should be capable of gracefully handling *exceptional* disturbances, i.e. disturbances of intensity higher than the predicted bounds which are used in the derivation of the control law. For instance, somebody may walk into the laboratory and push violently on the manipulator "to see how stiff it is"; an industrial robot may get jammed by the failure of some other machine; an actuator may saturate as the result of the specification of an unfeasible desired trajectory. In such cases the integral term in the control action may become unreasonably large, so that once the disturbance stops, the system goes through large amplitude oscillations in order to return to the desired trajectory. This phenomenon, known as *integrator windup*, is a potential cause of instability because of saturation effects and physical limits on the joint displacements. It can be simply avoided by *stopping integration* (i.e. maintaining the integral term in (6-52) constant) *as long as the system is outside the boundary layer*, for each degree of freedom i. Indeed, under normal circumstances the system does remain in the boundary layer; on the other hand, when the conditions return to normal after an exceptional disturbance, integration can resume (for each i) as soon as the system is back in the boundary layer, since the integral term in (6-52) is defined to within an arbitrary constant.

(iii) One can easily check that replacing (incorrectly) $k_i(\mathbf{q},\dot{\mathbf{q}})$ by $k_i(\mathbf{q}_d,\dot{\mathbf{q}}_d)$, in the computation of the term $\overline{k}_i(\mathbf{q},\dot{\mathbf{q}})$ used in the expression (6-51) of u_i, would make the control action in the boundary layer essentially identical to that of a pure computed torque scheme with integral control. However, of course, assuming that $k_i(\mathbf{q},\dot{\mathbf{q}}) = k_i(\mathbf{q}_d,\dot{\mathbf{q}}_d)$ would no longer guarantee that the trajectories actually *remain* in the boundary layer. Further, the use of a saturation function in sliding control, associated the boundary layer concept, prevents the control action from the potentially destabilizing effects of over-reacting to exceptional disturbances.

(iv) The implementation of a sliding controller only requires that the *diagonal* elements of $\mathbf{H}^{-1}\hat{\mathbf{H}}$ be positive in all configurations, as seen in (6-53). By contrast, it can be shown that a necessary condition for the stability of the pure computed torque scheme is that the *matrix* $\mathbf{H}^{-1}\hat{\mathbf{H}}$ be positive definite in all configurations, a much stronger constraint.[1]

(v) In the case that λ is time-varying (as further discussed in Section 6.5), the term

$$u_i' := -2\dot{\lambda}(\dot{\tilde{q}}_i + \lambda\tilde{q}_i)$$

should be added to control law u_i in (6-51), while augmenting gain $k_i(\mathbf{q},\dot{\mathbf{q}})$ of (6-57) accordingly by the quantity $|u_i'|(\beta_i - 1)$.

The degree of simplification in the system model may be varied according to the on-line computing power available: the balance conditions clearly quantify the trade-off between model precision and tracking accuracy as further detailed in Section 6.5. Further, the s-trajectories provide a measure of the validity of the assumptions on model uncertainty and of the adequacy of bound simplifications.

---

[1] Recall that the product of two positive definite matrices is not necessarily positive definite. Indeed, the scalar quantity $\mathbf{v}^T\mathbf{M}\mathbf{v}$ can be seen as the dot-product of the vector $\mathbf{v}$ with the transformed vector $\mathbf{M}\mathbf{v}$. Thus, for instance, the 2×2 matrix corresponding to a planar rotation of 60° is positive definite, while that corresponding to a rotation of 120° is not.

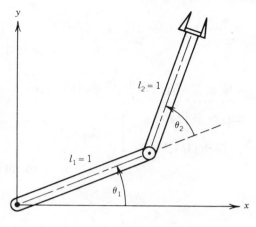

Figure 6-7 : Two-link manipulator.

Example 6-6: Planar Two-Link Manipulator

Consider the two-link planar manipulator of Figure 6-7. Both links are modeled to be of unit length and unit mass, although there is a 50% uncertainty on the mass of the second link. Thus the available model is, from (5-12):

$$
\hat{\mathbf{H}} = \begin{bmatrix} 5/3 + c_2 & 1/3 + 1/2 \, c_2 \\ 1/3 + 1/2 \, c_2 & 1/3 \end{bmatrix} = [\mathbf{H}_1 \ \ \mathbf{H}_2]
$$

$$
\hat{\mathbf{h}} = 1/2 \begin{bmatrix} -\dot{\theta}_2^2 - 2\dot{\theta}_1\dot{\theta}_2 \\ \dot{\theta}_1^2 \end{bmatrix} s_2
$$

(where $c_2 = cos[\theta_2]$, $s_2 = sin[\theta_2]$) while the actual values are of the form

$$
\mathbf{H} = \begin{bmatrix} 1/3 + (4/3 + c_2)a & (1/3 + 1/2 \, c_2)a \\ (1/3 + 1/2 \, c_2)a & a/3 \end{bmatrix} = [\mathbf{H}_1 \ \ \mathbf{H}_2]
$$

and

$$
\mathbf{h} = a\hat{\mathbf{h}} \tag{6-58}
$$

with

$$.5 \leq a \leq 1.5 \tag{6-59}$$

and assuming that friction is negligible. Letting $D := (a/9) + a^2(1/3 - c_2^2/4)$, we thus have

$$\mathbf{H}^{-1} = \frac{1}{D} \begin{bmatrix} a/3 & -(1/3 + 1/2\ c_2)a \\ -(1/3 + 1/2\ c_2)a & 1/3 + (4/3 + c_2)a \end{bmatrix} = [\mathbf{L}_1\ \mathbf{L}_2] \tag{6-60}$$

and

$$\mathbf{H}^{-1}\hat{\mathbf{H}} = \frac{1}{D} \begin{bmatrix} (4/9 - c_2^2/4)a & 0 \\ (1/9 + 1/6\ c_2)(1-a) & 1/9 + (1/3 - c_2^2/4)a \end{bmatrix} \tag{6-61}$$

$$= \begin{bmatrix} \mathbf{L}_1^T\hat{\mathbf{H}}_1 & \mathbf{L}_1^T\hat{\mathbf{H}}_2 \\ \mathbf{L}_2^T\hat{\mathbf{H}}_1 & \mathbf{L}_2^T\hat{\mathbf{H}}_2 \end{bmatrix}$$

One can easily write a computer program to systematically evaluate (6-57) as a function of θ . Actually, in this second-order example it would be somewhat tedious but reasonable to derive analytic expressions that tightly satisfy (6-57). For simplicity, we choose instead to verify (6-57) by using bounds slightly more conservative but quicker to evaluate. Namely, let us define the positive quantities

$$a_1 := \frac{1}{2}\ (4/9 - c_2^2/4)$$

$$a_2 := \frac{1}{2}\ (1/3 - c_2^2/4)$$

$$a_3 := \frac{1}{9} + a_2$$

$$a_4 := \frac{1}{9} + 3a_2$$

in order to account for the maximum variations of a in the numerators of (6-61), and

$$a_5 := \frac{1}{18} + \frac{1}{4}(1/3 - c_2^2/4) = \frac{1}{18} + \frac{a_2}{2}$$

$$a_6 := \frac{1}{6} + \frac{9}{4}(1/3 - c_2^2/4) = \frac{1}{6} + \frac{9a_2}{2}$$

to account for the variations of a in the denominator D of (6-61). We then let, according to (6-55) and (6-56)

$$\beta_1 := \left(\left(\frac{3a_1}{a_5}\right)\left(\frac{a_6}{a_1}\right)\right)^{1/2} = \left(\frac{3a_6}{a_5}\right)^{1/2}$$

$$\beta_2 := \left(\left(\frac{a_4}{a_5}\right)\left(\frac{a_6}{a_3}\right)\right)^{1/2} = \beta_1\left(\frac{a_4}{3a_3}\right)^{1/2}$$

$$G_1 := \left(\frac{a_1(3a_1)}{(a_5\,a_6)}\right)^{-1/2} = \frac{1}{a_1}\left(\frac{a_5\,a_6}{3}\right)^{1/2}$$

$$G_2 := \left(\frac{a_3\,a_4}{a_5\,a_6}\right)^{-1/2} = \left(\frac{a_5\,a_6}{a_3\,a_4}\right)^{1/2}$$

and from (6-58), (6-59)

$$\Delta \mathbf{h}^+ = \frac{1}{2}\,\hat{\mathbf{h}} \qquad\qquad (6-62)$$

These expressions and (6-60), (6-61) then allow one to compute $\mathbf{k}(\boldsymbol{\theta},\dot{\boldsymbol{\theta}})$ to satisfy (6-57).   ▲▲▲

6.5. The Modeling/Performance Trade-Offs

Conceptually, the development of Section 6.4 can be illustrated by Figure 6-8 : consider a fast manipulator motion, say a 1/2 second stop-to-stop trajectory across the workspace, including a full flipping of the wrist; and plot average tracking precision against parametric uncertainty (say average imprecision on manipulator inertias). Using the *computed torque method based on a full 6 d.o.f. model*, curve I is obtained: tracking is perfect in the absence of

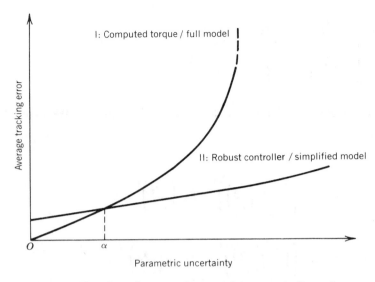

Figure 6-8 : The effect of parametric uncertainty on controller performance.

model uncertainty, and then quickly degrades as uncertainty increases, with the system eventually becoming unstable. Using a *robust controller based on a simplified model* (two "decoupled" 3 d.o.f. systems), the result is curve II: because of model simplification, performance is not "perfect" in the absence of parametric uncertainty, as when using the full 6 d.o.f. model; but the robust controller quickly outperforms the computed torque scheme as uncertainty increases. The crossing point α of Figure 6-8 typically corresponds to a very low level of parametric uncertainty; further, $\alpha \rightarrow 0$ as allowable bandwidth λ increases, as could be expected from the dynamic balance conditions of Section 6.3.3.   Robust control thus combines in practice improved performance with simpler and more tractable controller designs. Of course, the application of robust control to a full 6 d.o.f. model would yield a curve consistently below curve I .

The dynamic balance conditions (6-39)-(6-42) have important practical implications in terms of design/modeling/performance trade-offs. Neglecting time-constants of order $1/\lambda$, conditions (6-39) and (6-40) can be written

$$\lambda^n \epsilon \approx \beta_d k_d \tag{6-63}$$

as noticed in Section 6.3. One can easily show from (6-31) that, due to the presence of the integral control term in (6-52), relation (6-63) is in fact slightly modified here into (for each degree of freedom i)

$$\lambda^2 \epsilon \approx 2\beta_d k_d \qquad (6-64)$$

If we now consider the structure of control law (6-51), with $k_i(\mathbf{q},\dot{\mathbf{q}})$ obtained from (6-57), we see that the effects of parameter uncertainty on $\mathbf{H}$ and $\mathbf{h}$ have been "dumped" in gain $k_i(\mathbf{q},\dot{\mathbf{q}})$. Conversely, better knowledge of $\mathbf{H}$ or $\mathbf{h}$ reduces $k_i(\mathbf{q},\dot{\mathbf{q}})$ by a comparable quantity. Thus (6-64) is particularly useful in an *incremental* mode, i.e. to evaluate the effects of model simplification (or conversely of increased model sophistication) on tracking performance:

$$\Delta\epsilon \approx 2\Delta(\beta_d k_d/\lambda^2) \qquad (6-65)$$

In particular, marginal gains in performance are critically dependent on control bandwidth λ: if large λ's are available, poor dynamic models may lead to respectable tracking performance, and conversely large modeling efforts produce only minor absolute improvements in tracking accuracy. It is of course not overly surprising that system performance be very sensitive to control bandwidth λ : (6-1) only represents *part* of the system dynamics — namely, its rigid-body component — while λ accounts for the unmodeled part. Now, in robotics applications, λ is typically limited by three factors:

(i) *arm structural resonant modes*: λ must be smaller than the frequency ν_R of the lowest unmodeled structural resonant mode; a reasonable interpretation of this constraint is, classically (with λ_R in radians/second and ν_R in Hertz)

$$\lambda \leq \lambda_R \approx \frac{2\pi}{3}\nu_R \qquad (6-66)$$

although in practice this bound may be modulated by engineering judgment, taking notably into account the natural damping of the structural modes. Further, it may be worthwhile in certain cases to account for the fact that λ_R actually varies both with configuration and with loads.

(ii) *neglected time delays*: along the same lines, we have a condition of the form

$$\lambda \leq \lambda_A \approx \frac{1}{3T_A} \tag{6-67}$$

when T_A is the largest unmodeled time-delay (for instance in the actuators).

(iii) *sampling rate*: with a full-period processing delay, one gets a condition of the form

$$\lambda \leq \lambda_S \approx \frac{1}{5} \nu_{\text{sampling}} \tag{6-68}$$

where ν_{sampling} is the sampling rate.

Desired control bandwidth λ is the minimum of the three bounds (6-66)-(6-68). Bound (6-66) essentially depends on the arm's mechanical properties, while (6-67) reflects limitations on the actuators, and (6-68) accounts for the available computing power. Ideally, the most effective design corresponds to *matching* these limitations, i.e., having

$$\lambda_R \approx \lambda_A \approx \lambda_S =: \lambda \tag{6-69}$$

Now (6-66) and (6-67) are "hard" limitations, in the sense that they represent properties of the arm itself, while (6-68) is "soft" as far as it reflects the performance of the computer environment *and* the complexity of the control algorithm. Assume for instance that bound (6-68) is the most stringent, which means that the arm's mechanical potentials are not fully exploited. This may typically occur in modern high-performance robots (such as direct-drive arms) which feature high mechanical stiffness and high resonant frequencies. It may be worthwhile, before embarking in the development of dedicated computer architectures, to first consider *simplifying* the dynamic model used in the control algorithm. This in turn allows one to replace $\lambda = \lambda_{\text{slow}}$ by a larger $\lambda = \lambda_{\text{fast}}$ which varies inversely proportionally to the required computation time. From (6-64) and assuming that neither of bounds (6-66) or (6-67) is hit in the process, this operation is beneficial as long as

$$\frac{\Delta(\beta_d k_d)}{\beta_d k_d} \leq \left(\frac{\lambda_{fast}}{\lambda_{slow}}\right)^2 - 1 \tag{6-70}$$

Conversely, equality in (6-70) defines the threshold at which model simplification starts degrading performance despite gains in sampling rate. This threshold is rarely reached in practice: even assuming that marginal gains in model precision depend linearly on the computation time involved, λ^{-2} still varies as the square of the required sampling period. Thus it is often advisable to reduce model complexity until computation can be achieved at a rate fully compatible with the mechanical capabilities of the arm, in other words until λ_S is no longer the "active" limit on λ. The performance increase resulting from this simple operation may in turn be adequate and avoid major efforts on the computer hardware.

Similarly, simplifying the dynamic model may allow us to "push" bound (6-66). Consider again a 6 d.o.f. manipulator composed of a 3 d.o.f. "arm" and a 3 d.o.f. wrist. If we consider the whole 6x6 dynamics, λ is limited by the lowest resonant frequency of the whole manipulator. If instead we partition the computation into two 3x3 systems, regarded as disturbances to one another, the bandwidth at the wrist-level can be considerably increased. Again the simpler approach is actually beneficial (at least from the point of view of wrist precision) as long as (6-70) is satisfied - of course, further model simplifications at the wrist level may in turn be implied in order to match sampling rate to the new bandwidth available. Finally, if computation power is not an issue and the full model can be easily computed on-line, a convenient partitioning is to use the full 3x6 model to control the wrist actuators ($\lambda = \lambda_{fast}$), and a 3x3 model to control the "arm" actuators ($\lambda = \lambda_{slow}$).

Along similar lines, (6-64) and (6-66) imply that ultimately the cartesian-space tracking accuracy of a manipulator varies at least as the cube of its size. In other words, the improvement in cartesian accuracy to be expected from scaling down all dimensions of a manipulator by a given factor "size" varies at least as fast at $(size)^3$ — indeed, structural resonant frequencies vary at least as $(size)^{-1}$, as does the ratio of joint-space accuracy to cartesian-space accuracy. In high-precision applications (such as opto-electronic assembly), it is thus critical to use properly dimensioned robot arms, since the trade-off reach/accuracy is grossly biased toward small manipulators. Similarly, the variation of cartesian accuracy as $(size)^3$ emphasizes the practical importance of appropriate part feeding and conveying mechanisms: being able to design a robot twice as small by improving part-feeding increases tracking accuracy by a factor of eight.

Robust robot controller design provides a systematic approach to the problem of maintaining stability in the face of modeling imprecisions. Further, it quantifies the modeling/performance trade-offs, and in that sense illuminates the whole design and testing process. Finally it offers the potential of simplifying higher-level programming, by accepting reduced information about both task and manipulator.

6.6. Research Topics

Trajectory planning is concerned with the generation of desired trajectories that avoid hitting obstacles in the workspace and is traditionally treated as a geometry problem (e.g., Moravec, 1980; Lozano-Perez,1983; Brooks and Lozano-Perez, 1985). An additional degree of complexity is introduced by requiring that the task be completed in *minimum time*, given the physical *bounds on input torques* (e.g., Kahn and Roth, 1971; Gilbert and Johnson, 1985; Sahar and Hollerbach, 1985) — task execution time is, of course, directly related to productivity. The minimum-time control of a manipulator *along a given path* is addressed in (Bobrow, *et al.*, 1985). The above schemes are *open-loop*, i.e. they generate desired trajectories that do not account for model uncertainty — this implies that in practice the assumed bounds on input torques must be reduced in order to leave room for closed-loop control action. Alternatively, closed-loop pointwise optimal schemes are suggested in (Spong, Thorp, and Kleinwaks, 1984; Slotine and Spong, 1985). Also, the effective extension of trajectory planning schemes to redundant manipulators represents an important research area .

The problem of controlling robots with *flexible joints* has been studied by e.g. (Marino and Nicosia, 1984; Sweet and Good, 1985; Isidori *et al.*, 1985; Khorasani and Spong, 1985) and is based on the observation that most of the flexibility in "rigid" robots is concentrated at the joint level, as seen in Chapter 4. The problem of controlling robots with *flexible links* has been studied by e.g. (Cannon and Schmitz, 1984; Hastings and Book, 1985) and is motivated by the drive toward lighter and faster robots. The development of adequate sensors represents a large component of both problems (e.g. Nelson, 1986).

Finally, a number of researchers have studied the problem of reducing model uncertainty by using parameter estimation techniques. *Off-line* parameter estimation is discussed in e.g. (Atkeson, *et al.*, 1985; Khosla and Kanade, 1985).   *On-line* parameter estimation, adaptive control, and the closely related issue of improving the manipulator performance in repetitive tasks by introducing an initial "training" period, have been studied in e.g. (Dubowsky and Desforges, 1979; Horowitz and Tomizuka, 1980; Koivo and Guo, 1981; Leininger and Wang, 1982; Craig, 1984; Arimoto, 1985; Landau, 1985; Canudas *et al.*, 1986; Slotine and Coetsee, 1986).

NOTES AND REFERENCES

The stability proof of Section 6.2 is similar to that of (Arimoto and Miyazaki, 1983). The notion of sliding surface (Filippov, 1960) has been studied in great detail mostly in the Soviet literature (see Utkin, 1977 for a review).   Classical sliding mode control, however, presents important drawbacks, including chattering and large control authority, that limit its practical applicability.   An application of classical sliding mode control to robot manipulators is discussed in (Young, 1978).   Further details on the development of Sections 6.3 to 6.5 can be found in (Slotine, 1984, 1985).   Related approaches to robustness can be found in (Spong and Vidyasagar, 1985; Ha and Gilbert, 1985).   Effective algorithms to explicitly compute the inertia matrix and the Coriolis and centripetal torques are developed in (Walker and Orin, 1982), and use the Luh-Walker-Paul algorithm as as important component.

Appendix A.6.1: A Brush-Up on Control Theory

This Appendix briefly summarizes some of the key concepts of elementary control theory, which are abundantly illustrated in this chapter.   It is intended as a brushup rather than an introduction, for which the reader is referred to a good textbook on the subject, such as (Luenberger, 1979).

Given a system dynamics, the minimum information required to predict the system behavior for all future time is defined as the system *state*.   For a second-order system, for instance, the state can be expressed as position and velocity $\mathbf{X} := [x, \dot{x}]^T$.   The representation of the system state is not unique, since any one-to-one transformation of $\mathbf{X}$ is suitable.   Given the available system *input* u (say a force applied by an actuator), control consists in varying u with time so that the closed-loop dynamics have desirable properties, say that $\mathbf{X}$ follow a

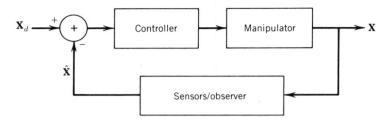

Figure A.6-1 : Feedback controller structure.

desired trajectory $\mathbf{X}_d$. *Feedback* control explicitly uses the measured error signal $\tilde{\mathbf{X}} = \mathbf{X} - \mathbf{X}_d$ as part of control law u , so as to reduce the system sensitivity to imprecision on the parameters used in the dynamic model (Figure A.6-1). However, besides involving stability issues, feedback control creates the risk of exciting high-frequency dynamics neglected in the system model. Thus, a desired control *bandwidth* is specified such that the frequency content of the control input be low enough not to excite the unmodelled dynamics. This in turn limits system performance, since in general the higher the bandwidth (i.e. the more "creative" the control action) the better the controller tracking precision.

Example 1: Second Order Linear System Response

For a second order system

$$\ddot{x} = u + d$$

where d is a disturbance, one can achieve a bandwidth of λ by letting

$$u := -2\eta\lambda\dot{\tilde{x}} - \lambda^2\tilde{x} \qquad\qquad (A.6-1)$$

The damping coefficient η can be adjusted to shape the transients appropriately, and is typically chosen so that $\eta \geq 0.7$. System responses to unit steps in d are plotted in Figure A.6-2 for typical values of damping coefficient η. Values of η larger or equal to one allow to avoid overshoot in response to step inputs in d. The value $\eta = 1$ is called *critical damping* and leads to the fastest overshot-free response.

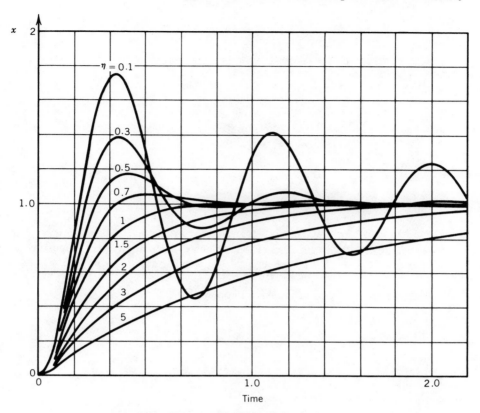

Figure A.6-2 : Step responses for typical values of the damping coefficient η.

For a desired constant position x_d, and a constant disturbance $d = d_o$, the resulting steady-state error is

$$\tilde{x} = \frac{d_o}{\lambda^2}$$

It is thus proportional to d_o and inversely proportional to the square of control bandwidth λ.

$$\boldsymbol{\Delta\Delta\Delta}$$

Example 2: Integral Control

The steady-state error can be eliminated by adding *integral control*, which makes the control action vary as long as $\tilde{\mathbf{X}}$ is not identically zero:

$$u = -3\lambda\dot{\tilde{x}} - 3\lambda^2\tilde{x} - \lambda^3\int_o^t \tilde{x}\,dT \qquad (A.6{-}2)$$

In steady-state we get

$$\int_o^t \tilde{x}\,dT \equiv \frac{d_o}{\lambda^3}$$

which implies

$$\tilde{x} \equiv 0$$

Such controller, which uses feedback in position, velocity, and the integral of position error, is often called a P.I.D. (proportional/integral/derivative) controller, and represents the single most widely used control structure in industrial applications. ◭◭◭

Example 3: Tracking

When the acceleration $\ddot{x}_d$ of the desired trajectory x_d is explicitly available, it is beneficial to include it in the control action (A.6-1)

$$u = \ddot{x}_d - 2\eta\lambda\dot{\tilde{x}} - \lambda^2\tilde{x}$$

Starting with arbitrary initial conditions, the tracking error $\tilde{x}$ then verifies

$$\ddot{\tilde{x}} + 2\eta\lambda\dot{\tilde{x}} + \lambda^2\tilde{x} = 0$$

and thus goes to zero within a few time constants λ^{-1} — the higher the bandwidth, the faster the convergence. Similarly, the term $\ddot{x}_d$ can be added to the P.I.D. control law (A.6-2).

◭◭◭

Lyapunov Functions

When the system dynamics is nonlinear

$$\dot{\mathbf{X}} = f(\mathbf{X})$$

a powerful tool for control system design is the notion of *Lyapunov function*. A Lyapunov function $V = V(\mathbf{X})$ can be thought of as a generalized energy function, and satisfies the following properties:

* V is a *scalar* function of the state $\mathbf{X}$ with continuous first partial derivatives.
* $\dot{V} \leq 0$ along all system trajectories.
* $V \to +\infty$ as $\|\mathbf{X}\| \to +\infty$

If we then define B as the region of the state space such that $\dot{V} = 0$, it can be shown that all system trajectories tend toward the largest *invariant set* within B. An invariant set is a region of the state-space such that all trajectories originating in this region remain in the region; typical examples are stable equilibrium points and limit cycles.

In the more general case that system dynamics depends not only on $\mathbf{X}$ but on external terms such as disturbances or desired trajectories:

$$\dot{\mathbf{X}} = \mathbf{f}(\mathbf{X};t)$$

the condition $\dot{V} \leq 0$ of the above development is replaced by

$$\dot{V} \leq -\eta\|\mathbf{X}\|$$

where η is some positive constant.

Example 4: Nonlinear Spring

Consider the nonlinear spring-damper system:

$$\ddot{x} + b(\dot{x}) + k(x) = 0$$

where b and k are continuous functions verifying

$$\dot{x}\, b(\dot{x}) > 0 \text{ for } \dot{x} \neq 0$$

$$x\, k(x) > 0 \text{ for } x \neq 0$$

which also imply that $b(0) = 0$ and $k(0) = 0$. A suitable Lyapunov function for this system is

$$V = \frac{1}{2}\dot{x}^2 + \int_0^x k(y)\ dy$$

which can be thought of as the sum of the kinetic and potential energy of the system. Indeed

$$\dot{V} = \dot{x}\ddot{x} + k(x)\dot{x}$$

$$= -\dot{x}b(\dot{x}) - k(x)\dot{x} + k(x)\dot{x}$$

$$= -\dot{x}b(\dot{x}) \leq 0$$

Further by hypothesis $\dot{x}b(\dot{x}) = 0$ only if $\dot{x} = 0$. Now $\dot{x} = 0$ implies that

$$\ddot{x} = -k(x)$$

which is nonzero as long as $x \neq 0$, so that $\dot{V}$ immediately becomes strictly negative again if $\dot{x} = 0$ and $x \neq 0$. Thus the system cannot get "stuck" at an equilibrium value other than $x = 0$; in other words, all trajectories converge towards $x = 0$.   **ΔΔΔ**

Example 5: Nonlinear Second Order System

Consider the problem of designing a controller to stabilize the system

$$\ddot{x} - \dot{x}^3 + x^2 = u$$

i.e. to bring it to equilibrium at $x \equiv 0$. Based on Example 4, it is sufficient to choose a continuous control law u of the form

$$u := u_1(\dot{x}) + u_2(x)$$

where

$$\dot{x}(\dot{x}^3 + u_1(\dot{x})) < 0' \quad \text{for} \quad \dot{x} \neq 0$$

$$x(x^2 - u_2(x)) > 0 \quad \text{for} \quad x \neq 0$$

For instance, the control law

$$u := -2\dot{x}^3 + x^2 - x$$

stabilizes the system. $\Delta\Delta\Delta$

Note that Lyapunov stability theory does not provide information on the frequency content of the control input. An approach to robustness to high-frequency unmodelled dynamics in nonlinear controller design is discussed in Section 6.3.

Example 6: Inverse Kinematics Algorithm

In Section 2.3.3, we saw that *analytical* inverse kinematic solutions exist only for special manipulator geometries. This represents a constraint on robot design, which may become important in certain applications such as deep sea or space robotics. Lyapunov stability theory can be used to obtain *computational* inverse kinematics algorithms that make no special assumption about geometry, and are also applicable to redundant manipulators.

Consider the problem of finding an inverse kinematics solution for a desired endpoint position $\mathbf{x}_d$. Let $\mathbf{q}$ be the current desired joint displacement estimate, and define the current error as

$$\tilde{\mathbf{x}} := \mathbf{x}(\mathbf{q}) - \mathbf{x}_d$$

Let us then select a Lyapunov function candidate as

$$V = \frac{1}{2} \tilde{\mathbf{x}}^T \mathbf{K}_P \tilde{\mathbf{x}} \qquad (A.6-3)$$

where $\mathbf{K}_P$ is a (not necessarily symmetric) positive definite matrix. Note that the function V is scalar, so that taking the transpose of (A.6-3) also implies

$$V = \frac{1}{2} \tilde{\mathbf{x}}^T \mathbf{K}_P^T \tilde{\mathbf{x}} \qquad (A.6-4)$$

Differentiating (A.6-4), we get

$$\dot{V} = \tilde{\mathbf{x}}^T \mathbf{K}_P^T \dot{\tilde{\mathbf{x}}} = \tilde{\mathbf{x}}^T \mathbf{K}_P^T \mathbf{J} \dot{\mathbf{q}}$$

so that defining the update law of estimate **q** according to

$$\dot{\mathbf{q}} := -\mathbf{J}^T \mathbf{K}_P \tilde{\mathbf{x}} \qquad (A.6-5)$$

we obtain

$$\dot{V} = -(\mathbf{J}^T \mathbf{K}_P \tilde{\mathbf{x}})^T (\mathbf{J}^T \mathbf{K}_P \tilde{\mathbf{x}}) = -\dot{\mathbf{q}}^T \dot{\mathbf{q}} \ \leq\ 0 \qquad (A.6-6)$$

The case $\dot{V} = 0$ corresponds to $\dot{\mathbf{q}} = \mathbf{0}$, i.e. from (A.6-5) to the case when $\mathbf{K}_P \tilde{\mathbf{x}}$ belongs to the null-space of $\mathbf{J}^T$. Hence a problem may arise if the current Jacobian matrix $\mathbf{J} = \mathbf{J}(\mathbf{q})$ is singular, since in that case the null-space of $\mathbf{J}^T$ does not reduce to the null vector, and thus $\tilde{\mathbf{x}}$ may be "stuck" at a non-zero value. This can be remedied in practice by adding to $\mathbf{K}_P$ an appropriate time-dependent *antisymmetric* matrix $K_A(t)$, which "rotates" the new K_P outside of the null space of $\mathbf{J}^T$ while leaving the function V of (A.6-4) unchanged. Indeed, for any vector **v**, and all t

$$\mathbf{v}^T \mathbf{K}_A(t)\mathbf{v} = (\mathbf{v}^T \mathbf{K}_A(t)\mathbf{v})^T = \mathbf{v}^T \mathbf{K}_A^T(t)\mathbf{v}$$

hence, since $\mathbf{K}_A(t)$ is antisymmetric (i.e., $\mathbf{K}_A^T(t) = -\mathbf{K}_A(t)$) :

$$\mathbf{v}^T \mathbf{K}_A(t)\mathbf{v} = -\mathbf{v}^T \mathbf{K}_A(t)\mathbf{v}$$

which in turn implies

$$\mathbf{v}^T \mathbf{K}_A(t)\mathbf{v} = 0$$

for all t, so that the addition of antisymmetric matrix $K_A(t)$ to K_P does not affect the Lyapunov function V of (A.6-4). A very similar situation will be encountered in Section 7.2.2 when we discuss impedance control. Note that the negative quantity $(\dot{V}/V) = (d/dt)\text{Log} \mid V\mid$, obtained directly from (A.6-3) and (A.6-6), represents a measure of how close the current configuration **q** is from being singular. **⊿⊿⊿**

Appendix A.6.2: Uncertainty on the Inertia Matrix

Although simplified conditions (6-57) are generally adequate in practice, exact expressions for the gains $k_i(\mathbf{q},\dot{\mathbf{q}})$ are presented here for completeness. Although somewhat involved mathematically, the derivation is of interest in itself since it provides insight on the effect of uncertainty about the manipulator mass properties.

Gains $k_i(\mathbf{q},\dot{\mathbf{q}})$ must actually verify, instead of the simplified condition (6-57)

$$k_i(\mathbf{q},\dot{\mathbf{q}}) \geq \beta_i \left(|\hat{u}_i| \, (1 - \beta_i^{-1}) + (\mathbf{L}_i^+)^T \, \Delta\mathbf{h}^+ + \sum_{j \neq i} \Delta_{ij} |u_j| + \eta_i \right)$$

$$(A.6-7)$$

where the scalars $\Delta_{ij} = \Delta_{ij}(\mathbf{q})$ are defined such that, for all $\mathbf{q}$

$$\Delta_{ij} \geq |\mathbf{L}_i^T \cdot \hat{\mathbf{H}}_j| \qquad i=1,\ldots,n \; ; \; j=1,\ldots,n$$

$$(A.6-8)$$

From the expression (6-51) of the u_j , and the definition (6-41) of the $\overline{k}_j(\mathbf{q},\dot{\mathbf{q}})$, a sufficient condition for (A.6-7) to be satisfied is that

$$k_i(\mathbf{q},\dot{\mathbf{q}}) = k'_i(\mathbf{q},\dot{\mathbf{q}}) + \beta_i \sum_{j \neq i} \left(G_j \, \Delta_{ij} \, k_j(\mathbf{q},\dot{\mathbf{q}}) |sat(s_j/\Phi_j)| \right)$$

$$(A.6-9)$$

where $k'_i(\mathbf{q},\dot{\mathbf{q}})$ is defined so that

$$k'_j(\mathbf{q},\dot{\mathbf{q}}) \geq \beta_j \left[|\hat{u}_i|(1 - \beta_i^{-1}) + (\mathbf{L}_i^+)^T \Delta\mathbf{h}^+ + \right.$$

$$\left. \sum_{j \neq i} G_j \, \Delta_{ij} \left| \hat{u}_j + [\lambda\Phi_j/\beta_j - k_j(\mathbf{q}_d,\dot{\mathbf{q}}_d)]sat(s_j/\Phi_j) \right| + \eta_i \right]$$

$$(A.6-10)$$

Now since all terms in $sat(s_j/\Phi_j)$ vanish for $\mathbf{q} \equiv \mathbf{q}_d$, $\dot{\mathbf{q}} \equiv \dot{\mathbf{q}}_d$, expressions (A.6-9) and (A.6-10) allow us to compute $k_i(\mathbf{q}_d,\dot{\mathbf{q}}_d) = k'_i(\mathbf{q}_d,\dot{\mathbf{q}}_d)$ directly. Further, from the balance conditions boundary layer thicknesses Φ_j only depend on the $k_j(\mathbf{q}_d,\dot{\mathbf{q}}_d)$ so that in turn the knowledge of the $k_j(\mathbf{q}_d,\dot{\mathbf{q}}_d)$ allows (A.6-9) and (A.6-10) to completely define the $k_j(\mathbf{q},\dot{\mathbf{q}})$: the vector $\mathbf{k}(\mathbf{q},\dot{\mathbf{q}})$ of components $k_j(\mathbf{q},\dot{\mathbf{q}})$ is the solution of the linear system

$$A(\mathbf{q},\dot{\mathbf{q}})\mathbf{k}(\mathbf{q},\dot{\mathbf{q}}) = \mathbf{k}'(\mathbf{q},\dot{\mathbf{q}}) \tag{A.6-11}$$

where $\mathbf{k}'(\mathbf{q},\dot{\mathbf{q}})$ is the vector of components $k_i'(\mathbf{q},\dot{\mathbf{q}})$, and the matrix $\mathbf{A}(\mathbf{q},\dot{\mathbf{q}})$ is defined as

$$\mathbf{A}(\mathbf{q},\dot{\mathbf{q}}) := \begin{bmatrix} 1 & -\beta_1 G_2 \Delta_{12}|sat(s_2/\Phi_2)| & \cdots \\ -\beta_2 G_1 \Delta_{21}|sat(s_1/\Phi_1)| & 1 & \\ \vdots & & \ddots \end{bmatrix}$$

Further, all components $k_i(\mathbf{q},\dot{\mathbf{q}})$ of the solution $\mathbf{k}(\mathbf{q},\dot{\mathbf{q}})$ of (A.6-11) must be strictly positive for this solution to be admissible: given the Metzler structure (Siljak, 1978) of the matrix $-\mathbf{A}$, this is the case as long as all real eigenvalues of $\mathbf{A}$ remain positive. Equivalently, defining $\mathbf{A}_o = \mathbf{A}_o(\mathbf{q},\dot{\mathbf{q}})$ as

$$\mathbf{A}_o := \mathbf{diag}(\beta_j^{-1})\mathbf{A}\ \mathbf{diag}(G_j^{-1})$$

$$= \begin{bmatrix} \beta_1^{min} & -\Delta_{12}|sat(s_2/\Phi_2)| & \cdots \\ -\Delta_{21}|sat(s_1/\Phi_1)| & \beta_2^{min} & \\ \vdots & & \ddots \end{bmatrix}$$

(where the second equality stems from definitions (6-54) and (6-55) of the G_j and β_j), $\mathbf{k}(\mathbf{q},\dot{\mathbf{q}})$ of (A.6-11) is admissible as long as all real eigenvalues of $\mathbf{A}_o$ remain positive. Note that $\mathbf{A} \equiv \mathbf{A}_o \equiv I$ in the absence of uncertainty on the inertia matrix.

Remarks

(i) A sufficient condition for $\mathbf{k}(\mathbf{q},\dot{\mathbf{q}})$ to be admissible is that all real eigenvalues of the matrix

$$\mathbf{A}'_o := \begin{bmatrix} \beta_1^{min} & -\Delta_{12} & \cdots \\ -\Delta_{21} & \beta_2^{min} & \\ \vdots & & \ddots \end{bmatrix}$$

be positive.

(ii) Regardless of the level of parametric uncertainty, it is *always* possible to generate admissible $\mathbf{k}(\mathbf{q},\dot{\mathbf{q}})$ by selecting large enough $\mathbf{k}'(\mathbf{q},\dot{\mathbf{q}})$. This can be achieved by multiplying the right-hand side of inequality (A.6-10) by an appropriately large scaling factor ρ , such as a constant upper bound on the Frobenius-Perron roof[1] ρ_1 of the matrix

$$\mathbf{A}_1 := \begin{bmatrix} 0 & \beta_1 G_2 \Delta_{12} & \cdots \\ \beta_2 G_1 \Delta_{21} & 0 & \\ \vdots & & \ddots \end{bmatrix}$$

(where we assume that $\rho_1 > 1$, else gain vector $\mathbf{k}(\mathbf{q},\dot{\mathbf{q}})$ is already guaranteed to be admissible). This has the effect of artificially increasing boundary layer thicknesses Φ_j by the same scaling factor without actually modifying the control action inside the (original and still effective) boundary layers, since gains $\overline{k}_j(\mathbf{q},\dot{\mathbf{q}})$ also scale up by the same factor ρ . The values of the s_j are thus left unchanged (as is the tracking performance, which allows us to use constant ρ's), so that the off-diagonal terms of the matrix $\mathbf{A}$ are divided by the scaling factor, which in turn guarantees that the solution $\mathbf{k}(\mathbf{q},\dot{\mathbf{q}})$ of (A.6-11) is admissible. Further, it is desirable that (A.6-11) be guaranteed to have admissible solutions regardless of the values of the s_j . This can be achieved by substituting $\rho_I^{-1} sat(\rho_I s_j / \Phi_j)$ to $sat(s_j / \Phi_j)$ in the expression of $\mathbf{A}$, with $\rho_1 \leq \rho_I \leq \rho$; which further guarantees that the B_j^o are attractive for all trajectories starting

---

[1]See Siljak, 1978. The Frobenius-Perron roof ρ_1 of a matrix $\mathbf{A}_1$ with non-negative elements is the largest (automatically non-negative) real eigenvalue of $\mathbf{A}_1$. The equation $(\mathbf{I} - \mathbf{A}_1/\rho)\mathbf{y} = \mathbf{x}$ admits component-wise positive solutions $\mathbf{y}$ for component-wise positive $\mathbf{x}$ if $\rho > \rho_1$.

inside boundary layers of thicknesses $(\rho/\rho_1)\Phi_j^o$, where Φ_j^o is the thickness of B_j^o. Note that a convenient upper bound of ρ_1 is

$$\rho_1 \leq \left(\max_i \left(G_i \sum_{j \neq i} \Delta_{ji}\beta_j \right) , \max_i \left(\beta_i \sum_{j \neq i} \Delta_{ij}G_j \right) \right)$$

(iii) The solution $\mathbf{k}(\mathbf{q},\dot{\mathbf{q}})$ of (A.6-11) is bounded for bounded $\mathbf{q}_d$, provided all real eigenvalues of $\mathbf{A}_o$ (or $\mathbf{A}$) are *uniformly* bounded away from zero, i.e. remain larger than some strictly positive constant. Indeed, $\mathbf{A}^{-1}$ ($\mathbf{q}$, $\dot{\mathbf{q}}$) is then bounded for bounded $\mathbf{q}$ and $\dot{\mathbf{q}}$ and further $\mathbf{k}(\mathbf{q},\dot{\mathbf{q}})$ is admissible; this in turn implies that tracking error is indeed limited by boundary layer thicknesses Φ_j, which depend only on the $k_j(\mathbf{q}_d,\dot{\mathbf{q}}_d)$, so that $\mathbf{q}$, $\dot{\mathbf{q}}$ and thus both $\mathbf{A}^{-1}(\mathbf{q},\dot{\mathbf{q}})$ and $\mathbf{k}'(\mathbf{q} ,\dot{\mathbf{q}})$ are bounded.

(iv) The condition that $\mathbf{k}(\mathbf{q},\dot{\mathbf{q}})$ of (A.6-11)be admissible can be given a slightly different interpretation (based again on the Metzler structure of $-\mathbf{A}_o$): it is satisfied if

$$\det \begin{bmatrix} \beta_1^{min} & -\Delta_{12}x_2 & \cdots \\ -\Delta_{21}x_1 & \beta_2^{min} & \\ \vdots & & \ddots \end{bmatrix} > 0 \qquad (A.6-12)$$

for all χ_j such that $0 \leq \chi_j \leq |sat(s_j/\Phi_j)|$. Let us now assume that

$$\det \hat{\mathbf{H}} > 0 \qquad (A.6-13)$$

Condition (A.6-13) is automatically satisfied for any physically motivated choice of $\hat{\mathbf{H}}$, since then $\hat{\mathbf{H}} = \hat{\mathbf{H}}^T > \mathbf{0}$. Now from $\mathbf{H} = \mathbf{H}^T > \mathbf{0}$, (A.6-13) implies that

$$\det(\mathbf{H}^{-1}\hat{\mathbf{H}}) > 0 \qquad (A.6-14)$$

which can be written

$$
\det \begin{bmatrix} \mathbf{L}_1^T \hat{\mathbf{H}}_1 & \mathbf{L}_1^T \hat{\mathbf{H}}_2 & \cdots \\ \mathbf{L}_2^T \hat{\mathbf{H}}_1 & \mathbf{L}_2^T \hat{\mathbf{H}}_2 & \\ \vdots & & \ddots \end{bmatrix} > 0
\qquad (A.6-15)
$$

The comparison of (A.6-12) and (A.6-15) shows that it is desirable that the $\beta_i^{\min}$ of (6-54) and the Δ_{ij} of (A.6-7) be compatible, i.e. be generated in a way that preserves the natural structure of (A.6-15).

Chapter 7
COMPLIANT MOTION CONTROL

7.1. Introduction

Most of today's industrial robots are used for spray-painting, pick-and-place or spot welding, all of which operations can generally be adequately handled by simple position control loops. Demanding tasks such as plasma welding or laser cutting require more sophisticated trajectory control capabilities. Yet very few manipulators are capable of such seemingly simple tasks as driving a screw or turning a crank. Human-like operations such as manipulating fragile objects or assembling toy-cars are barely in the reach of the most advanced experimental robots. Robotic assembly is still much more a laboratory favorite than a widespread industrial reality.

Compliant motion control is concerned with the control of a robot in contact with its "environment" — an object to manipulate or assemble, a welding seam to follow, and so on — and represents a major topic of current research. Consider for instance the problem of writing "Compliant Motion Control" on a blackboard. Pure trajectory control is clearly not recommended, since minute errors along the direction orthogonal to the board may result in not touching the board, or on the contrary in breaking the chalk. While trajectory control is needed along the board plane, so that the "text" be readable, some kind of *force control* is required in the direction orthogonal to the board, so as to guarantee that adequate contact is maintained. More formally, we can define a *compliance frame* (or *task frame*) in which the task to be performed is easily described, as illustrated in Figure 7-1. A compliance frame is a time-varying orthogonal coordinate system, such that at each instant and along each axis the task can be expressed as a pure trajectory *or* force (torque) control problem. The "or" is of course exclusive: one cannot control a single degree of freedom in both position and force, no more

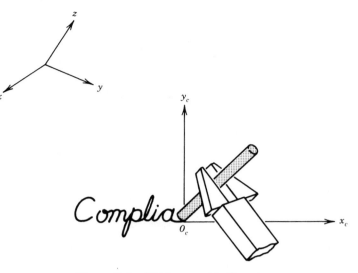

Figure 7-1 : Writing on a blackboard.

than one can specify *both* voltage and current across a resistor. In our "writing" example
(Figure 7-1), the manipulator endpoint first moves towards the blackboard with six d.o.f. in
position/orientation, and no freedom in force/torque since no source of reaction force is
available. At the instant when the chalk reaches the board, one d.o.f. in position is lost, and
accordingly one d.o.f. in force is gained along the z_c axis. If the chalk was glued to the board,
there would be no freedom in position/orientation, but six d.o.f. in force/torque. Conversely,
if the blackboard was replaced by a "soft" surface one could also choose to control position
instead of force along the z_c axis.

The duality between position and force control can be expressed in terms of *natural
constraints* and *artificial constraints*. Natural constraints in position or force are defined by the
geometry of the task to be performed. In our writing example, for instance, the presence of the
blackboard represents a natural constraint in position along the z_c direction. Moreover, if the
contact between the chalk and the blackboard is assumed to be frictionless, then two natural
force constraints arise along the x_c and y_c directions since the tangential force must be zero.
Three additional natural constraints express the fact that no reaction torque is available
at O_c, so that torques about the x_c , y_c and z_c axes of the compliance frame must be zero.
Further, the desired trajectories or forces specified by the user define "artificial" constraints

associated with the task. In our writing example, the artificial constraints consist of the text to be written (trajectory constraints in x_c and y_c) and of the desired orientation of the end-effector. The artificial constraints must be compatible with the natural constraints since again one cannot control both force and position along a given degree of freedom. In other words, position and force along each degree of freedom are each determined *either* by a natural constraint *or* an artificial constraint. Thus the number of natural constraints and the number of artificial constraints are both equal to the number of degrees of freedom of the constraint space (6 in general). Note that in certain cases some of the natural constraints may be expressed in terms of the artificial constraints, or vice-versa. For instance, if Coulomb friction is used to model the contact between the chalk and the blackboard, then the tangential forces along x_c and y_c depend on the normal force along z_c .

In our writing example, the compliance frame origin O_c is chosen to be the contact point between the chalk and the blackboard. The desired task can then be described as obtaining a desired contact force (or a range of admissible contact forces) in the z_c direction, while maintaining coordinates x_c and y_c at zero. In other words the desired trajectory (here the "text") is described by specifying the motion of the compliance frame origin O_c . More generally, the choice of the compliance frame depends on the specific task to be performed. Examples of typical tasks and possible choices of compliance frames are illustrated in Figure 7-2 , along with their associated natural and artificial constraints. Vector coordinates expressed in the compliance frame are referred to as *task-space coordinates*.

Two approaches to compliant motion control are detailed in this chapter: *impedance control* and *dynamic hybrid control*. Impedance Control, discussed in Section 7.2, does not specify desired forces or positions, but rather desired dynamic relationships *between* force and position, in other words mechanical stiffnesses or "impedances". It can be seen as an extension of the local schemes of Section 6.2, and presents very similar characteristics: it is extremely simple and robust to parameter uncertainty, but is mostly restricted to fairly slow motions. Dynamic hybrid control is discussed in Section 7.3, and is the compliant-motion version of the computed torque techniques encountered in trajectory control. It consists in rewriting manipulator dynamics directly in terms of the end-effector coordinates (or *operational* coordinates) instead of the joint displacements $\mathbf{q}$, and then controlling position, force, or

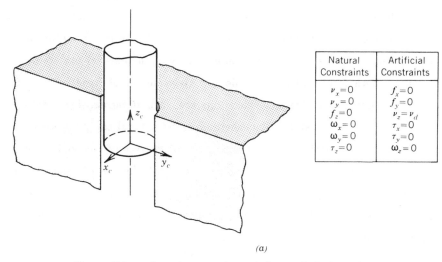

The table in figure (a):

Natural Constraints	Artificial Constraints
$\nu_x = 0$	$f_x = 0$
$\nu_y = 0$	$f_y = 0$
$f_z = 0$	$\nu_z = \nu_d$
$\omega_x = 0$	$\tau_x = 0$
$\omega_y = 0$	$\tau_y = 0$
$\tau_z = 0$	$\omega_z = 0$

(a)

Figure 7-2.a : Inserting a peg into a hole at a desired speed v_{d}.

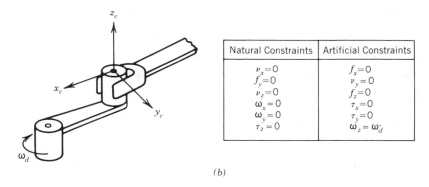

Natural Constraints	Artificial Constraints
$\nu_x = 0$	$f_x = 0$
$f_y = 0$	$\nu_y = 0$
$\nu_z = 0$	$f_z = 0$
$\omega_x = 0$	$\tau_x = 0$
$\omega_y = 0$	$\tau_y = 0$
$\tau_z = 0$	$\omega_z = \omega_d$

(b)

Figure 7-2.b : Turning a crank at a desired angular velocity ω_{d}.

mechanical impedance along each axis of the compliance frame. Although it does account for full manipulator dynamics at all speeds, pure dynamic hybrid control presents robustness problems similar to those of the computed torque method, only compounded by the fact that dynamics is generally significantly more involved in task-space than in joint-space. It is thus mostly suitable for robots with simple dynamics, or at fairly low speeds where the dynamics is dominated by gravity and friction. More generally, compliant motion control presents specific robustness problems linked to the strong effects of stiction, joint and link flexibility, and transmission backlash. Robustness issues, further discussed in Section 7.4, represent an active area of current research.

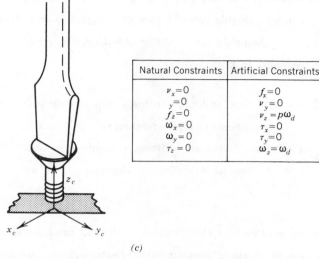

Natural Constraints	Artificial Constraints
$\nu_x=0$	$f_x=0$
$\nu_y=0$	$\nu_y=0$
$f_z=0$	$\nu_z=p\omega_d$
$\omega_x=0$	$\tau_x=0$
$\omega_y=0$	$\tau_y=0$
$\tau_z=0$	$\omega_z=\omega_d$

(c)

Figure 7-2.c. : Driving a screw of pitch p at a desired angular velocity ω_d.

7.2. Impedance Control

7.2.1. Passive and Active Compliance

To solve the specific "writing" problem of the previous section, it is conceivable to equip the end-effector with a *passive* mechanical device composed of springs and dampers, that the robot would maintain in an appropriate orientation. Similarly, (Drake, 1977) shows that peg-in-hole insertions can be facilitated by appropriately introducing low lateral and rotational stiffnesses in the grasping mechanism. A passive mechanical device known as the RCC (Remote Center Compliance) is based on this principle, and exploits the observation that it is easier to "pull" an object into a hole than to "push" it into the hole (Whitney,1982). In more technical terms, the RCC allows one to place the *compliance center* at the tip of the peg. A compliance center is a point such that a force applied at that point causes a pure translation, while a pure torque applied at the point causes a pure rotation about the point. When a compliance center exists (or is artificially created), it represents a natural choice for the compliance frame origin O_c .

Passive mechanical devices such as the RCC are typically capable of quick responses, and are relatively inexpensive. However, their application is necessarily limited to very specific tasks — the RCC, for instance, can only handle pegs of a certain length and orientation with respect to the hand. A programmable *active* device, by contrast, would allow the manipulation of various types of parts, or the modification of the end-effector's elastic behavior according to different phases of an assembly task.

In section 4.2, we saw that endpoint stiffness depends on joint-servo stiffness, joint mechanical compliance, and (in general to a lesser extent) on link flexibility. Conversely, as we shall now see, we can compute the joint stiffnesses required in order to obtain a desired endpoint stiffness. These desired joint stiffnesses can in turn be achieved by appropriate controller design.

In this section, we assume that the effects of both joint mechanical compliance and link flexibility can be neglected — robustness to such effects will be discussed in Section 7.4. Consider then the problem of achieving a desired, programmable stiffness of the end-effector. The desired elastic behavior can be described by a *stiffness matrix* $\mathbf{K}_P$, expressed in task-space. Namely, the desired restoring force $\mathbf{F}$, to be generated at the end-effector in response to small displacements $\delta\mathbf{x}$ from a nominally commanded position $\mathbf{x}_d$, is defined as

$$\mathbf{F} = -\mathbf{K}_P\,\delta\mathbf{x} \tag{7-1}$$

where $\mathbf{F}$, $\mathbf{K}_P$ and $\delta\mathbf{x}$ are all expressed in task-space coordinates. The positive definite matrix $\mathbf{K}_P$ is generally chosen to be diagonal, and composed of desired low stiffnesses for task-space directions along which stiffness must be controlled, and of large numbers (only limited by the available control bandwidth) for the remaining task-space directions, along which position must be controlled. Now the desired restoring force (7-1) can be actually achieved by applying joint torques

$$\tau = \mathbf{J}^T\mathbf{F} \tag{7-2}$$

where $\mathbf{J}$ is the manipulator Jacobian. The Jacobian is also expressed in task-space coordinates, and relates virtual end-effector displacements to virtual joint displacements:

$$\delta\mathbf{x} = \mathbf{J}\delta\mathbf{q} \tag{7-3}$$

so that (7-1) and (7-2) can be rewritten as

$$\boldsymbol{\tau} = -(\mathbf{J}^T \mathbf{K}_P \mathbf{J}) \delta \mathbf{q} \qquad\qquad (7-4)$$

The term $\mathbf{K_q} := \mathbf{J}^T \mathbf{K}_P \mathbf{J}$ in equation (7-4) is called the *joint stiffness matrix*. It allows one to simply express the task-space stiffness specification (7-1) in terms of joint torques $\boldsymbol{\tau}$ and joints displacements $\mathbf{q}$, i.e. of the variables most directly relevant to the control system. The joint stiffness matrix $\mathbf{K_q}$ is not diagonal, in other words arbitrary endpoint stiffnesses can only be achieved by suitably *coordinating* joint torques in response to small endpoint displacements. Also, $\mathbf{K_q}$ is degenerate at manipulator singularities, which implies that active stiffness control is not possible in certain directions. This is not surprising since we know that at singularities the manipulator cannot move in all directions, nor can it exert forces in all directions.

The Jacobian matrix in (7-3) and (7-4) can be computed at any point of the end-effector (or even at any point fixed with respect to the end-effector). Thus we can not only specify the orthogonal directions (or *principal stiffness directions*) along which given stiffnesses must be achieved, but also actively position the compliance center anywhere in the end-effector. This capability is extremely useful in assembly, since it allows us both to arbitrarily move the compliance center (which can then be chosen as the origin of the compliance frame), and to specify the principal stiffness directions (which can then coincide with the axes of the compliance frame) and the corresponding desired stiffnesses, according to the different phases of the assembly task.

7.2.2. Active Impedance Control

Impedance control is an extension of the above development to arbitrarily large task-space displacements:

$$\tilde{\mathbf{x}} := \mathbf{x} - \mathbf{x}_d$$

It consists in controlling the dynamic interactions between the manipulator and its environment directly in task-space by applying the control law

$$\tau = \hat{\mathbf{g}}(\mathbf{q}) - \mathbf{J}^T(\mathbf{q})[\mathbf{K}_P \tilde{\mathbf{x}} + \mathbf{K}_D \dot{\tilde{\mathbf{x}}}] \tag{7-5}$$

where $\hat{\mathbf{g}}(\mathbf{q})$ is the estimated gravitational torque, and $\mathbf{J}(\mathbf{q})$ is the manipulator Jacobian. Displacement vector $\tilde{\mathbf{x}}$ and matrices $\mathbf{J}^T$ and $\mathbf{K}_P$ are generally expressed directly in task-space coordinates. Matrices $\mathbf{K}_P$ and $\mathbf{K}_D$ in (7-5) can be interpreted as the desired apparent "stiffness" and "damping" of the manipulator as seen from the environment. The task-space force $-[\mathbf{K}_P \tilde{\mathbf{x}} + \mathbf{K}_D \dot{\tilde{\mathbf{x}}}]$ is then transformed into a joint torque vector through the use of $\mathbf{J}^T(\mathbf{q})$, according to (4-7). We see from expression (7-5) that, by definition, impedance control places the compliance center at the reference position $\mathbf{x}_d$.

Impedance control owes its name to the fact that (7-5) monitors the dynamic relationship *between* force and position, rather than directly controlling force or position. Further, the comparison of (7-5) with (6-9) shows that impedance control can also be seen as an extension of the local control philosophy of Section 6.2. As such, we can expect it to feature similar advantages, namely simplicity and robustness to parametric uncertainty (except for the effects of gravity); and the same drawbacks, in particular limited dynamic performance. We now show that this is indeed the case, but that in addition the scheme presents some problems in the vicinity of manipulator singularities, as can be expected from the presence of the Jacobian matrix $\mathbf{J}(\mathbf{q})$ in expression (7-5). In Section 7.2.2.1 we first consider whether control law (7-5) makes sense as a *position* control scheme, before discussing the properties of (7-5) as a *compliance* control scheme in Section 7.2.2.2.

7.2.2.1. Impedance Control as a Position Control Scheme

Consider the problem of freely moving a manipulator endpoint to a specified position $\mathbf{x}_d$, fixed in cartesian space. Similarly to the development of Section 6.2, we can define the Lyapunov function candidate

$$V := \frac{1}{2} [\tilde{\mathbf{x}}^T \mathbf{K}_P \tilde{\mathbf{x}} + \dot{\mathbf{q}}^T \mathbf{H} \dot{\mathbf{q}}] \tag{7-6}$$

which can again be interpreted as the total energy associated with the closed-loop system. We assume again that the gravitational component is exactly compensated, i.e. $\hat{\mathbf{g}}(\mathbf{q}) = \mathbf{g}(\mathbf{q})$. Differentiating (7-6), and noting that $\dot{\mathbf{H}}$ simplifies with the matrix $\mathbf{C}$ as in Proof 6.2 of Section

6.2, we get

$$\dot{V} = \dot{\mathbf{x}}^T \mathbf{K}_P \tilde{\mathbf{x}} - \dot{\mathbf{q}}^T \mathbf{J}^T (\mathbf{K}_P \tilde{\mathbf{x}} + \mathbf{K}_D \dot{\tilde{\mathbf{x}}})$$

Noting that $\dot{\tilde{\mathbf{x}}} = \dot{\mathbf{x}} = \mathbf{J}\dot{\mathbf{q}}$ then leads to

$$\dot{V} = -\dot{\mathbf{x}}^T \mathbf{K}_D \dot{\mathbf{x}} \leq 0 \qquad\qquad (7\text{--}7)$$

and thus shows that control strategy (7-5) is stable. To determine whether it is *asymptotically* stable, i.e. whether it actually brings the manipulator endpoint to $\mathbf{x}_d$, we must (as in Proof 6.2) analyze the case $\dot{V} = 0$, i.e., from (7-7), the case $\dot{\mathbf{x}} = \mathbf{0}$. Now

$$\dot{\mathbf{x}} = \mathbf{0} \qquad => \qquad \ddot{\mathbf{q}} = -\mathbf{H}^{-1}\mathbf{J}^T\mathbf{K}_P\tilde{\mathbf{x}} - \mathbf{H}^{-1}\mathbf{C}\dot{\mathbf{q}} \qquad\qquad (7\text{--}8)$$

so that three alternatives can occur:

(i) The manipulator is not redundant, and $\mathbf{J}(\mathbf{q})$ has full rank at the current arm configuration $\mathbf{q}$. Then $\dot{\mathbf{x}} = \mathbf{0}$ implies that $\dot{\mathbf{q}} = \mathbf{0}$, so that (7-8) yields

$$\ddot{\mathbf{q}} = -\mathbf{H}^{-1}\mathbf{J}^T\mathbf{K}_P\tilde{\mathbf{x}}$$

which is indeed non-zero as long as $\tilde{\mathbf{x}} \neq \mathbf{0}$, since $\mathbf{J}$ and thus $\mathbf{H}^{-1}\mathbf{J}^T\mathbf{K}_P$ are non-singular.

(ii) The Jacobian matrix $\mathbf{J}(\mathbf{q})$ is degenerate at the current $\mathbf{q}$ (in other words, the current manipulator configuration is singular). The situation is much less clear then, since no conclusion on $\tilde{\mathbf{x}}$ can be drawn directly from equation (7-8). In particular, there is a whole range of manipulator *motions* such that $\mathbf{J}(\mathbf{q})$ *remains* singular — for the two-link articulated manipulator of Figure 7-3, for instance, such is the case for rotations about the origin ($\dot{q}_1 \neq 0$) with the arm completely stretched ($q_2 \equiv 0$) or completely folded ($q_2 \equiv \pi$). The problem is, however, not as bad as it looks, since in practice there always is a small amount of viscous friction $\mathbf{D}\dot{\mathbf{q}}$ added to dynamics (6-1), with $\mathbf{D}$ a positive definite (generally diagonal) matrix — one can in fact purposefully add such a term to the control torque $\boldsymbol{\tau}$. Expression (7-7) then becomes

$$\dot{V} = -\dot{\mathbf{x}}^T\mathbf{K}_D\dot{\mathbf{x}} - \dot{\mathbf{q}}^T\mathbf{D}\dot{\mathbf{q}} \leq -\dot{\mathbf{q}}^T\mathbf{D}\dot{\mathbf{q}} \leq 0 \qquad\qquad (7\text{--}9)$$

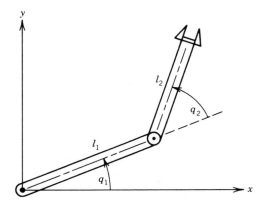

Figure 7-3 : Two-link manipulator.

so that $\dot{V} = 0$ does imply that $\dot{\mathbf{q}} = \mathbf{0}$, hence that

$$\ddot{\mathbf{q}} = -\mathbf{H}^{-1}\mathbf{J}^T\mathbf{K}_P\tilde{\mathbf{x}}$$

Now since $\mathbf{J}(\mathbf{q})$ is degenerate at the current $\mathbf{q}$, there exist non-zero values of $\tilde{\mathbf{x}}$ such that $\mathbf{K}_P\tilde{\mathbf{x}}$ belong to the null space of $\mathbf{J}^T$. The impedance control law (7-5) "stalls" at such values. Physically, this corresponds to requiring that the manipulator actively apply a force $\mathbf{K}_P\tilde{\mathbf{x}}$ in a direction along which it cannot move. Consider for instance the two-link manipulator of Figure 7-3, with $\mathbf{K}_P = \mathbf{I}$, and assume that the task-frame coincides with the cartesian reference frame. The Jacobian matrix

$$\mathbf{J} = \begin{bmatrix} -l_1 s_1 - l_2 s_{12} & l_1 c_1 + l_2 c_{12} \\ -l_2 s_{12} & l_2 c_{12} \end{bmatrix}$$

(where $c_i := \cos(q_i)$, $s_i := \sin(q_i)$, $c_{ij} := \cos(q_i + q_j)$, and $s_{ij} := \sin(q_i + q_j)$) is singular at $q_2 = 0$ (arm stretched) and $q_2 = \pi$ (arm folded). For the singularity $q_2 = 0$, we get

$$\mathbf{J}^T = \begin{bmatrix} -(l_1 + l_2)s_1 & (l_1 + l_2)c_1 \\ -l_2 s_1 & l_2 c_2 \end{bmatrix}$$

and thus $\boldsymbol{\tau} = -\mathbf{J}^T\mathbf{K}_P\tilde{\mathbf{x}} = \mathbf{0}$ if $s_1\tilde{x}_1 = c_1\tilde{x}_2$, i.e. if $\mathbf{x}_d$ is aligned with the arm.   More generally, for $\mathbf{K}_P \neq \mathbf{I}$, there is a line of $\mathbf{x}_d$ such that $\boldsymbol{\tau} = \mathbf{0}$ while $\tilde{\mathbf{x}} \neq \mathbf{0}$, as illustrated in Figure 7-4. A similar result is obtained for the singularity $q_2 = \pi$.

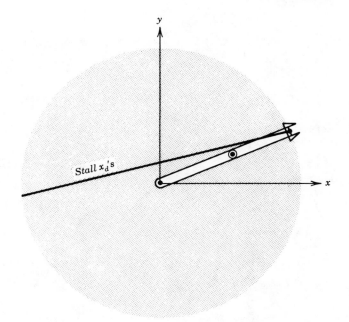

Figure 7-4 : Stall values of the reference position $\mathbf{x}_d$.

(iii) The manipulator is redundant, but $\mathbf{J(q)}$ does have full row-rank at the current $\mathbf{q}$. The inclusion of the viscous friction term $\mathbf{D\dot{q}}$ as in case (ii) then guarantees that the manipulator will not get stuck at the current $\mathbf{q}$.

The potential "stall" effect of case (ii) may be remedied in practice by adding to $\mathbf{K}_P$ a time-varying (e.g. state-dependent) *antisymmetric* matrix $\mathbf{K}_A(t)$. Adding $\mathbf{K}_A(t)$ to $\mathbf{K}_P$ leaves the Lyapunov function V of (7-6) unchanged and thus does not affect stability, but "rotates" the new $\mathbf{K}_P \tilde{\mathbf{x}}$ out of the null-space of $\mathbf{J}^T$ for appropriate choices of $\mathbf{K}_A(t)$, in a fashion similar to Example 6 of Appendix A.6.1. Note that the new $\mathbf{K}_P$ does not correspond to any "physical" stiffness matrix, since a passive stiffness matrix is necessarily symmetric (Prigogine, 1967).

REMARK: Impedance control does not require us to explicitly solve the manipulator inverse kinematics, since (7-5) is expressed directly in terms of the cartesian error vector $\tilde{\mathbf{x}}$. Conversely, impedance control can itself be used as a *computational scheme* for solving inverse kinematics problems, in a fashion very similar to Example 6 of Appendix A.6.1. To this effect, it suffices to simulate the impedance control law (7-5) as applied to a

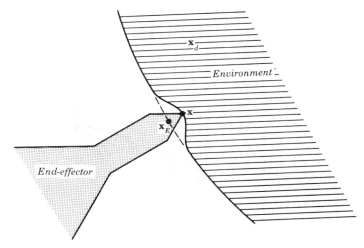

Figure 7-5 : Elastic model (7-10).

hypothetical robot of the very same geometry as that of the manipulator of interest, but with simple dynamics such that $\mathbf{H} = \mathbf{I}$ and $\mathbf{h} = \mathbf{0}$. The inverse kinematics solution estimate $\mathbf{q}$ is thus generated iteratively according to the update law

$$\ddot{\mathbf{q}} + \mathbf{D}\dot{\mathbf{q}} + \mathbf{J}^T(\mathbf{K}_P\tilde{\mathbf{x}} + \mathbf{K}_D\dot{\tilde{\mathbf{x}}}) = \mathbf{0}$$

where

$$\tilde{\mathbf{x}} := \mathbf{x}(\mathbf{q}) - \mathbf{x}_{\mathbf{d}}$$

The positive definite matrices $\mathbf{D}$, $\mathbf{K}_P$ and $\mathbf{K}_D$ can then be used to shape the transients, i.e. to accelerate the algorithm convergence and select among multiple inverse kinematics solutions. Note that the algorithm can be directly applied to redundant manipulators.

7.2.2.2. Impedance Control as a Compliance Control Scheme

Consider now the manipulator once it is in contact with its environment (Figure 7-5). The local deformation of the environment due to contact can be represented by a vector $\tilde{\mathbf{x}}_E$

$$\tilde{\mathbf{x}}_E = \mathbf{x} - \mathbf{x}_E \quad \text{when in contact}$$
$$\tilde{\mathbf{x}}_E = \mathbf{0} \qquad \text{otherwise}$$

The associated reaction force $\mathbf{F}_E$ exerted by the environment on the manipulator can then be modeled as an elastic restoring force

$$\mathbf{F}_E = -\mathbf{K}_E \tilde{\mathbf{x}}_E \qquad\qquad (7-10)$$

where the positive definite matrix $\mathbf{K}_E$ describes the environment stiffness. The vector $\mathbf{x}_E$ can be seen as representing the location to which the contact point $\mathbf{x}$ would return in the absence of control torque or gravity. In order to maintain contact at rest, the reference endpoint position $\mathbf{x}_d$ must be "inside" the environment (as illustrated in Figure 7-5) since it represents the rest position of the spring-damper system defined by matrices $\mathbf{K}_P$ and $\mathbf{K}_D$. Remark that environment stiffness matrix $\mathbf{K}_E$ and deformation $\tilde{\mathbf{x}}_E$ are convenient but idealized notions: it is more accurate to consider contact force $\mathbf{F}_E$ of (7-10) as the result of the *collective* deformation of the environment *and* the manipulator. Further, note that friction is neglected in expression (7-10).

The stiffness matrix $\mathbf{K}_P$ of control law (7-5) is selected according to the desired compliant task, similarly to what was done in Section 7.2.1. For instance one can choose $\mathbf{K}_P$ to be diagonal, and composed of desired low stiffnesses for directions along which force interactions must be monitored, and of large elements (only limited by the available control bandwidth) for directions along which trajectory must be controlled. The choice of $\mathbf{K}_P$ may also be adjusted to account for the flexibility of the manipulator itself. The matrix $\mathbf{K}_D$ can then be chosen accordingly to be diagonal and composed of the desired damping coefficients in each direction, with critical damping selected along trajectory-controlled directions.

To examine the stability of impedance control, for a fixed $\mathbf{x}_d$, we again select our Lyapunov function candidate as the total system energy:

$$V := \frac{1}{2}\left[\tilde{\mathbf{x}}^T \mathbf{K}_P \tilde{\mathbf{x}} + \dot{\mathbf{q}}^T \mathbf{H} \dot{\mathbf{q}} + \tilde{\mathbf{x}}_E^T \mathbf{K}_E \tilde{\mathbf{x}}_E\right]$$

The additional term $\tilde{\mathbf{x}}_E^T \mathbf{K}_E \tilde{\mathbf{x}}_E$ in the expression of V, as compared with definition (7-6), accounts for the potential energy associated with the elastic interaction (7-10) between the manipulator and the environment. The system dynamics is now

$$\mathbf{H\ddot{q}} + \mathbf{C\dot{q}} + \mathbf{g(q)} = \boldsymbol{\tau} + \mathbf{J}^T\mathbf{F}_E = \boldsymbol{\tau} - \mathbf{J}^T\mathbf{K}_E\,\tilde{\mathbf{x}}_E$$

Noting that conservation of energy implies $\mathbf{F}_E^T \cdot \dot{\mathbf{x}}_E = 0$, we obtain

$$\dot{V} \leq -\dot{\mathbf{q}}^T\mathbf{D\dot{q}} \leq 0$$

similarly to (7-9). Thus again $\dot{V} = 0$ only if $\dot{\mathbf{q}} = \mathbf{0}$, with now

$$\dot{\mathbf{q}} = \mathbf{0} \quad => \quad \ddot{\mathbf{q}} = -\mathbf{H}^{-1}\mathbf{J}^T(\mathbf{K}_P\tilde{\mathbf{x}} + \mathbf{K}_E\tilde{\mathbf{x}}_E)$$

Assuming that we are not at a manipulator singularity, the equilibrium point $\mathbf{x}$ thus corresponds to

$$\mathbf{K}_P\,\tilde{\mathbf{x}} + \mathbf{K}_E\,\tilde{\mathbf{x}}_E = \mathbf{0} \qquad\qquad (7-11)$$

i.e. is given by the weighted average

$$\mathbf{x} = (\mathbf{K}_P + \mathbf{K}_E)^{-1}\,(\mathbf{K}_P\mathbf{x}_d + \mathbf{K}_E\mathbf{x}_E) \qquad\qquad (7-12)$$

which reflects the combined effects of the environment stiffness and the desired manipulator impedance. From (7-11) the corresponding value of the Lyapunov function V is

$$V = (\mathbf{x}_E - \mathbf{x}_d)^T\,\mathbf{K}_P\,(\mathbf{x} - \mathbf{x}_d)$$

which we can rewrite from (7-12) as a quadratic form in $(\mathbf{x}_E - \mathbf{x}_d)$:

$$V = (\mathbf{x}_E - \mathbf{x}_d)^T\,\mathbf{K}_P\,(\mathbf{K}_P + \mathbf{K}_E)^{-1}\mathbf{K}_E\,(\mathbf{x}_E - \mathbf{x}_d)$$

At singularities, the manipulator may again get "stuck" at values of $\mathbf{x}$ different from (7-12), but such that the current $(\mathbf{K}_P\tilde{\mathbf{x}} + \mathbf{K}_E\,\tilde{\mathbf{x}}_E)$ belongs to the null-space of $\mathbf{J}^T(\mathbf{q})$. This may again be remedied by adding to $\mathbf{K}_P$ an appropriate state-dependent antisymmetric matrix.

It is important to remark that the above development does not require interaction force measurements or explicit knowledge of the environment stiffness matrix $\mathbf{K}_E$. In other words, the scheme is robust no only to parametric uncertainty on the manipulator itself, but also to imprecision on the manipulator environment. Further, impedance control law (7-5) can be directly applied to redundant manipulators, as already noticed in Section 7.2.2.1.

7.3. Dynamic Hybrid Control

Similarly to the case of the local P.I.D. schemes of Section 6.2, an approach more sophisticated than impedance control is required to achieve effective compliant *motion* control, and fully account for system dynamics. The associated theory, introduced in this section, is not yet quite as mature as that of pure trajectory control, but constitutes a major topic of current research since a thorough understanding of compliant motion control represents a key element toward effective industrial robotics.

Section 7.3.1 first discusses force control in the simple one-dimensional case. In Section 7.3.2, the development is extended to the dynamic hybrid control of cartesian manipulators: trajectory is controlled along certain directions of the compliance frame, while force (or impedance) is monitored along the other directions. The hybrid control framework is generalized to arbitrary manipulators in Section 7.3.3, by rewriting manipulator dynamics in terms of end-effector coordinates.

7.3.1. Force Control

Consider the single degree-of-freedom system

$$m\ddot{x} = F + F_E + d \tag{7-13}$$

where F is the control force exerted by the actuator, while

$$F_E = -k_E \, \tilde{x}_E \tag{7-14}$$

is the reaction force exerted by the environment (similarly to (7-10)), and d is a disturbance force (due for instance to friction, backlash, and so on). The spring constant k_E is typically of the order of 10^6 N/m for a stiff environment. The control problem is to maintain F_E to a desired value F_d, with the corresponding error signal given by

$$\tilde{F} := F_E - F_d \qquad\qquad (7-15)$$

Contrary to the impedance control scheme of the previous section, we must now *assume that F_E is available for measurement*. The force sensors in a robot may be colocated with some actuator, or may be placed at the fingertips, or in the manipulator environment itself, or (most commonly) at the manipulator wrist. Depending on the sensor location, the determination of the actual interaction forces may require a geometric transformation (as illustrated in Example 4-2), as well as the calculation of inertial acceleration effects if the arm is in motion. Further, force measurements are often fairly noisy, mostly because of friction and other imperfections in the transmission mechanisms.

Note that *human* dexterity strongly relies strongly on force information. This can be easily understood by trying to write while wearing mittens, for example. Similarly, the addition of force feedback to early teleoperators considerably improved the performance of the global man-machine system.

Given (7-14), the differentiation of (7-15) leads to

$$\ddot{\tilde{F}} = \ddot{F}_E - \ddot{F}_d = -k_E \ddot{\tilde{x}}_E - \ddot{F}_d$$

Assuming a fixed environment, we thus get, using dynamics (7-13)

$$\ddot{\tilde{F}} = -k_E m^{-1}(F + F_E + d) - \ddot{F}_d \qquad\qquad (7-16)$$

In the expressions above, we assume that the environment stiffness k_E of (7-19) is exactly known. This assumption further allows us to differentiate (7-14) to get

$$\dot{F}_E = -k_E \dot{x} \qquad\qquad (7-17)$$

Although based on an approximation, expression (7-17) generally leads to better estimates of

$\dot{F}_E$ than those that would be obtained from differentiating a noisy force signal F_E. We can now choose control law F of (7-16) so that $\tilde{F}$ verifies a desired closed loop dynamics:

$$\ddot{\tilde{F}} := -\alpha_D \dot{\tilde{F}} - \alpha_P \tilde{F} - k_E m^{-1} d \qquad (7-18)$$

where the term due to disturbance d is derived from (7-16). Equating the right-hand sides of (7-16) and (7-18) leads to the desired control action:

$$F := -F_E + m k_E^{-1} (-\ddot{F}_d + \alpha_D \dot{\tilde{F}} + \alpha_P \tilde{F}) \qquad (7-19)$$

Note from (7-18) that for a given bandwidth (reflected in the choice of α_P and α_D), the force error $\tilde{F}$ varies in inverse proportion to the system mass m, but proportionally to the environment stiffness k_E. These dependences both spring from the fact that the force dynamic model (7-16) used to determine control action is actually based on position dynamics (7-13) through the use of elastic model (7-14).

This single d.o.f. example already illustrates the importance of robustness issues. In particular, the development is critically dependent on a good estimate of environment stiffness k_E, since k_E is used both in the derivation of the model (7-16) of force dynamics and in (7-17) to compute the state feedback component $\dot{F}_E$ in order to obtain the desired damping term α_D $\dot{\tilde{F}}$. What this means in practice is that a force-controlled manipulator may exhibit a perfectly reasonable behavior when you push it with your hand, yet become unstable if you push it using a steel rod. Indeed, one sees from (7-19) that underestimating environment stiffnes k_E leads to a higher effective bandwidth, hence to potential unstability. On the other hand, grossly overestimating k_E would lead to a very conservative bandwidth, hence to sluggish response and high sensitivity to disturbances.

As already noticed in Section 7.2.2, the "environment" stiffness k_E is actually associated with the elastic deformation of the environment *and* the manipulator. When a force sensor is used, as is the case here, the term k_E thus also incorporates the stiffness of the sensor itself. It may thus seem beneficial to use a force sensor with a low stiffness, that then dominates the stiffness of the environment. However, a sensor of low stiffness actually filters the force signal, thus degrading the quality of the force information and the capacity of the controller to effectively react to fast changes in contact force (as those that occur when making contact with the environment, or while following an irregular surface, for instance). Further, using a sensor of very low stiffness may generate uncertainty about the actual position of the endpoint when the manipulator is *not* in contact with the environment, or more generally in the directions along which position is to be controlled. Conversely, these effects are reduced by having a smaller inertia "beyond" the force sensor, i.e. by sensing force at the fingertip rather than at the wrist. Such engineering trade-offs represent an important part of any practical implementation of force sensing.

Robustness issues will be further discussed in Section 7.4.

7.3.2. Hybrid Control of a Cartesian Manipulator

Consider the two-dimensional cartesian manipulator of Figure 7-6:

$$\mathbf{H}\ddot{\mathbf{x}} = \mathbf{F} + \mathbf{F}_E + \mathbf{d}$$

where the inertia matrix $\mathbf{H}$ is constant and diagonal. The manipulator is required to follow the environment surface (Σ) at a given speed v_d while maintaining a desired contact force F_d. Contact between the manipulator endpoint and (Σ) is assumed to be frictionless, so that the interaction force is normal to the surface (Σ).

The control problem is "hybrid", in the sense that a desired trajectory (determined by (Σ) and v_d) is specified along the x_c axis of the compliance frame, while the contact force $\mathbf{F}_E$ is to be monitored about the desired value $\mathbf{F}_d$ along the y_c axis. Thus, defining $\mathbf{n}_{cx}$ as the unit vector along the x_c axis, the only part of $\mathbf{x}$ of interest for trajectory control is the *projection* of $\mathbf{x}$ onto $\mathbf{n}_{cx}$, namely $\mathbf{n}_{cx}(\mathbf{n}_{cx}^T \mathbf{x})$. Similarly, defining $\mathbf{n}_{cy}$ as the unit vector along the y_c axis, the only part of $\mathbf{F}_E$ of interest for force control is the projection of $\mathbf{F}_E$ on $\mathbf{n}_{cy}$, namely $\mathbf{n}_{cy}(\mathbf{n}_{cy}^T \mathbf{F}_E)$. Equivalently, these projections can be computed by first expressing $\mathbf{x}$ and $\mathbf{F}_E$ in task-space coordinates, then *selecting* the component of interest in trajectory or force control, and finally transforming back into the original cartesian coordinate system:

$$\mathbf{n}_{cx}\,\mathbf{n}_{cx}^T = \mathbf{R} \begin{pmatrix} 1 & 0 \\ 0 & 0 \end{pmatrix} \mathbf{R}^{-1} \qquad\qquad (7\!-\!20)$$

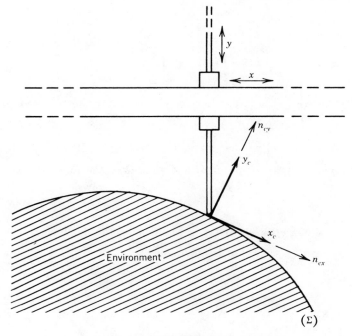

Figure 7-6 : A hybrid control problem.

$$\mathbf{n_{cy}}\ \mathbf{n_{cy}^T} = \mathbf{R} \begin{pmatrix} 0 & 0 \\ 0 & 1 \end{pmatrix} \mathbf{R}^{-1} \qquad (7{-}21)$$

where $\mathbf{R}$ is the rotation matrix:

$$\mathbf{R} := [\mathbf{n_{cx}}\ \mathbf{n_{cy}}]$$

which reflects the desired coordinate transformation. Equations (7-20) and (7-21) merely amount to expressing desired task-space projection matrices in the original cartesian coordinate system. Further, $\mathbf{R}^{-1}$ is simply $\mathbf{R}^T$, since $\mathbf{R}$ is a rotation matrix. Expressions (7-20) and (7-21) can be rewritten using a *selection vector* $\boldsymbol{\sigma} = [1\ 0]^T$, which specifies that the x_c component is to be trajectory-controlled, while the y_c component is to be force-controlled:

$$\mathbf{n_{cx}}\ \mathbf{n_{cx}^T} = \mathbf{R}\ \mathbf{diag}(\boldsymbol{\sigma})\ \mathbf{R}^T \qquad (7{-}22)$$

$$\mathbf{n_{cy}}\ \mathbf{n_{cy}^T} = \mathbf{R}\ (\mathbf{I} - \mathbf{diag}(\boldsymbol{\sigma}))\ \mathbf{R}^T \qquad (7{-}23)$$

where

$$\mathbf{diag}(\pmb{\sigma}) := \begin{bmatrix} \sigma_1 & 0 \\ 0 & \sigma_2 \end{bmatrix} = \begin{bmatrix} 1 & 0 \\ 0 & 0 \end{bmatrix}$$

Pure trajectory or force control laws are then applied to the projected error signals, and are then *added* to obtain the final hybrid control law, as summarized in Figure 7-7. Combining these two control laws is justified by the fact that projections (7-22) and (7-23) guarantee that the manipulator motion is not overspecified, and thus that the force and trajectory control laws are compatible.

Note the presence of inertia matrix $\mathbf{H}$ in Figure 7-7 : the dynamic model used to design the controller actually rewrites the robot free dynamics

$$\mathbf{H\ddot{x}} = \mathbf{F}$$

in the form

$$\mathbf{\ddot{x}} = \mathbf{u} \quad ; \quad \mathbf{F} = \mathbf{Hu}$$

This manipulation is purely notational, since $\mathbf{H}$ is diagonal for a cartesian robot, but it will prove to be convenient in the next section when we extend the hybrid control framework to general manipulators. Also, by convention, the unspecified task-space components of desired trajectory or force are set to zero in the implementation of Figure 7-7. Similarly, one only needs to measure interaction force (or position) along the directions where force (or trajectory) is to be controlled.

Example 7-1

If, for instance, the surface (Σ) of Figure 7-6 was defined parametrically as

$$x_\Sigma = x_\Sigma(\xi)$$
$$y_\Sigma = y_\Sigma(\xi)$$

with the desired tangential motion specified by selecting $\xi := \xi(t)$, the coordinate

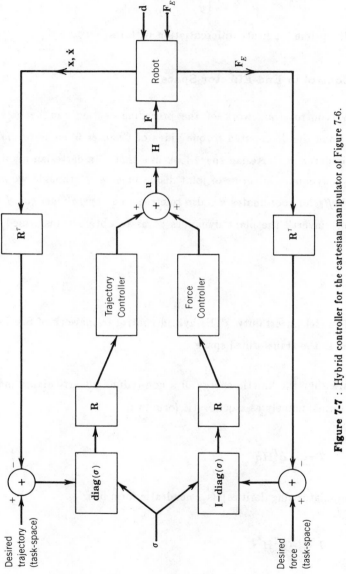

Figure 7-7 : Hybrid controller for the cartesian manipulator of Figure 7-6.

transformation $\mathbf{R}$ would be given by

$$\mathbf{R} := (x_\Sigma'^2 + y_\Sigma'^2)^{-1/2} \begin{bmatrix} x'_\Sigma & -y'_\Sigma \\ y'_\Sigma & x'_\Sigma \end{bmatrix}$$

where superscripts "prime" indicate differentiation with respect to ξ .   ▵▵▵

7.3.3. Hybrid Control in End-Effector Space

The hybrid control framework of the preceding section can be extended to general manipulators by using the "computed torque" idea of Chapter 6, so as to make the dynamics of a general manipulator in Cartesian space look like that of a cartesian manipulator. Instead of expressing robot dynamics in terms of joint displacements $\mathbf{q}$, though, we must now rewrite it in terms of *end-effector* coordinates $\mathbf{x}$, also referred to as *operational coordinates*. This then allows us to again "invert" the plant dynamics so as to obtain transformed dynamics of the simple form

$$\ddot{\mathbf{x}} = \mathbf{u} \tag{7-24}$$

in the absence of model uncertainty. The hybrid control framework of Section 7.3.2 can then be directly applied in the transformed space.

We first remark that the kinetic energy of a non-redundant articulated manipulator can be expressed in two forms, namely as a quadratic form in $\dot{\mathbf{q}}$

$$T = \frac{1}{2} \dot{\mathbf{q}}^T \mathbf{H} \dot{\mathbf{q}} \tag{7-25}$$

or (except at manipulator singularities) as a quadratic form in $\dot{\mathbf{x}}$

$$T = \frac{1}{2} \dot{\mathbf{x}}^T \mathbf{H}^* \dot{\mathbf{x}} \tag{7-26}$$

The form (7-25) was used in Section 5.2.3 to derive manipulator dynamics in terms of joint displacements $\mathbf{q}$. Alternatively, we can use form (7-26) to express the dynamics as a function of operational coordinates $\mathbf{x}$, similarly to the discussion of Section 5.2.4. Lagrange equations are then written, with $L := T - U$, as

$$\frac{d}{dt}\left(\frac{\partial L}{\partial \dot{\mathbf{x}}}\right) - \frac{\partial L}{\partial \mathbf{x}} = \mathbf{F} \qquad (7-27)$$

where $\mathbf{F}$ is the "operational force", which can be regarded as the force to apply at the end-effector in order to obtain the motion $\mathbf{x}(t)$. From the principle of virtual work, the same motion $\mathbf{x}(t)$ can be obtained, instead, by applying joint torque vector

$$\boldsymbol{\tau} := \mathbf{J}^T(\mathbf{q})\,\mathbf{F} \qquad (7-28)$$

The comparison of expressions (7-25) and (7-26), given that

$$\dot{\mathbf{x}} = \mathbf{J}\dot{\mathbf{q}}$$

leads to

$$\mathbf{H}^* = \mathbf{J}^{-T}\,\mathbf{H}\,\mathbf{J}^{-1} \qquad (7-29)$$

outside of manipulator singularities. Further, equation (7-27) can be written

$$\mathbf{H}^*\ddot{\mathbf{x}} + \dot{\mathbf{H}}^*\dot{\mathbf{x}} + \mathbf{v} - \frac{\partial U}{\partial \mathbf{x}} = \mathbf{F} \qquad (7-30)$$

where the components v_i of vector $\mathbf{v}$ are defined as

$$v_i = -\frac{1}{2}\dot{\mathbf{x}}^T\left(\frac{\partial L}{\partial x_i}\right)\dot{\mathbf{x}}$$

Using expression (7-29) in equation (7-30), and comparing with joint-space dynamics (6-1), one gets after some manipulation

$$\mathbf{H}^*\ddot{\mathbf{x}} + \mathbf{h}^* = \mathbf{F} \qquad (7-31)$$

where $\mathbf{H}^*$ is defined by (7-29) and

$$\mathbf{h}^* = \mathbf{J}^{-T}\mathbf{h} - \mathbf{H}^*\dot{\mathbf{J}}\dot{\mathbf{q}} \qquad (7-32)$$

System dynamics can thus be written in the simple form (7-24) by letting

$$\mathbf{F} := \mathbf{H}^*\mathbf{u} + \mathbf{h}^*$$

that is, given transformation (7-28):

$$\tau := \mathbf{J}^T\mathbf{F} = (\mathbf{H}\mathbf{J}^{-1})\,(\mathbf{u} - \dot{\mathbf{J}}\dot{\mathbf{q}}) + \mathbf{h} \qquad (7-33)$$

Note that expression (7-33) can also be obtained formally from (6-1) and $\dot{\mathbf{x}} = \mathbf{J}\dot{\mathbf{q}}$. Of course, the torque actually applied is

$$\tau := (\hat{\mathbf{H}}\mathbf{J}^{-1})\,(\mathbf{u} - \dot{\mathbf{J}}\dot{\mathbf{q}}) + \hat{\mathbf{h}} \qquad (7-34)$$

instead of (7-33) (assuming that the manipulator geometry is precisely known), so that robustness problems similar to those of the computed torque method have to be properly accounted for, as further detailed in Section 7.4.

The cartesian hybrid control structure of Figure 7-7 can now be directly extended to general manipulators by making manipulator dynamics transparent to the hybrid combination process, as illustrated in Figure 7.8. Again, the selection vector σ is composed of zeroes and ones, and specifies whether the corresponding compliance frame coordinate is to be controlled in force or in position. Trajectory and force controllers are then designed independently based on the appropriate projections of trajectory or force errors onto the compliance frame axes, and are combined into the hybrid control vector $\mathbf{u}$ associated to transformed dynamics (7-24). Finally, the corresponding joint torque vector τ to be applied is computed from $\mathbf{u}$ according to (7-34). Note that the terms involved in the transformation (7-34) from $\mathbf{u}$ to τ can generally be computed at a slower update rate than those composing control law $\mathbf{u}$ itself. Also, note that $\mathbf{R}$ in Figure 7-8 is not necessarily a rotation matrix, but more generally any coordinate transformation between cartesian space and task-space, so that in general $\mathbf{R}^{-1}$ does not equal $\mathbf{R}^T$.

Example 7-2

The coordinate transformations in dynamic hybrid control can be considerably simplified by a judicious choice of representation. Let $\mathbf{F}_d$ and τ_d be the two 3×1 vectors of

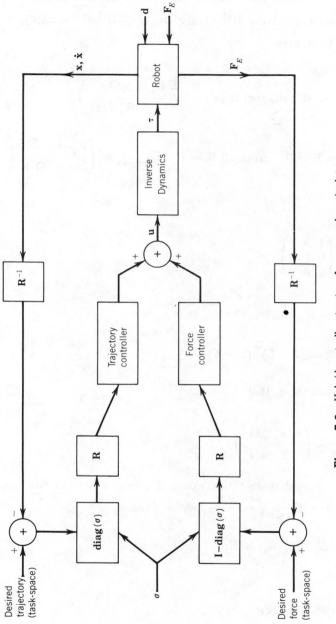

Figure 7-8 : Hybrid controller structure for a general manipulator.

desired force and torque at the end-effector, and let $\mathbf{R}_F$ and $\mathbf{R}_\tau$ be the 3×3 rotation matrices that would bring the z axis of cartesian reference frame $O\text{-}xyz$ in alignment with $\mathbf{F}_d$ and $\boldsymbol{\tau}_d$, respectively. The matrix products $\mathbf{R^{-1}\ diag(\sigma)\ R}$ and $\mathbf{R^{-1}\ [I - diag(\sigma)]\ R}$ that appear in Figure 7-8 can then be written

$$\boldsymbol{\Sigma}_{\text{traject}} := \mathbf{R^{-1}\ diag(\sigma)\ R} = \begin{pmatrix} \mathbf{R}_F^T\,\boldsymbol{\Sigma}\,\mathbf{R}_F & \mathbf{O} \\ \mathbf{O} & \mathbf{R}_\tau^T\,\boldsymbol{\Sigma}\,\mathbf{R}_\tau \end{pmatrix}$$

$$\boldsymbol{\Sigma}_{\text{force}} := \mathbf{R}^{-1}\,[\mathbf{I} - \mathbf{diag(\sigma)}]\,\mathbf{R} = \mathbf{I} - \boldsymbol{\Sigma}_{\text{traject}} = \begin{pmatrix} \mathbf{R}_F^T(\mathbf{I} - \boldsymbol{\Sigma})\mathbf{R}_F & \mathbf{O} \\ \mathbf{O} & \mathbf{R}_\tau^T(\mathbf{I} - \boldsymbol{\Sigma})\mathbf{R}_\tau \end{pmatrix}$$

where

$$\boldsymbol{\Sigma} := \begin{pmatrix} 1 & 0 & 0 \\ 0 & 1 & 0 \\ 0 & 0 & 0 \end{pmatrix}$$

from the definitions of $\mathbf{R}_F$ and $\mathbf{R}_\tau$. Noting that

$$(\mathbf{I} - \boldsymbol{\Sigma}) = (\mathbf{I} - \boldsymbol{\Sigma})^T\,(\mathbf{I} - \boldsymbol{\Sigma})$$

a quick calculation then shows that

$$\boldsymbol{\Sigma}_{\text{force}} = \begin{pmatrix} \mathbf{e}_F\,\mathbf{e}_F^T & \mathbf{O} \\ \mathbf{O} & \mathbf{e}_\tau\mathbf{e}_\tau^T \end{pmatrix}$$

where $\mathbf{e}_F^T$ is the last line of the rotation matrix $\mathbf{R}_F$ and $\mathbf{e}_\tau^T$ is the last line of the rotation matrix $\mathbf{R}_\tau$. The matrix $\boldsymbol{\Sigma}_{\text{traject}}$ can then be computed as

$$\boldsymbol{\Sigma}_{\text{traject}} = \mathbf{I} - \boldsymbol{\Sigma}_{\text{force}} \qquad\qquad \Delta\Delta\Delta$$

7.4. Robustness Issues

In this section we further discuss the robustness issues that arise in the compliant motion control of manipulators due to the presence of parametric uncertainty and disturbances. We first consider the case of dynamic hybrid control in section 7.4.1, and then take a closer look at robustness in impedance control, in section 7.4.2.

7.4.1. Robustness in Dynamic Hybrid Control

We first detail the one-dimensional force control problem of Section 7.3.1, and then generalize our results. Consider again the single degree-of-freedom system (7-13)

$$m\ddot{x} = F + F_E + d \tag{7-35}$$

required to maintain a desired contact force F_E with its environment. We noticed in Section 7.3.1 that the performance of basic controller (7-19) was potentially very sensitive to parametric uncertainty and in particular to imprecision on the environment stiffness k_E of (7-14). To address robustness issues in more detail, we can again use the notations and concepts of sliding control introduced in Section 6.3. Assume for instance that k_E is only known to within a factor 2 :

$$\frac{1}{2} \leq \frac{\hat{k}_E}{k_E} \leq 2 \tag{7-36}$$

while the estimate $\hat{m}$ of the manipulator mass satisfies

$$\frac{2}{3} \leq \frac{\hat{m}}{m} \leq \frac{3}{2}$$

and assume further that disturbance force d (due for instance to friction on backlash) is bounded by a known (perhaps time-varying) quantity:

$$|d| \leq D$$

We can then express the problem of maintaining $\tilde{F} = F_E - F_d$ at zero in terms of tracking the sliding surface

$$s = \dot{\tilde{F}} + 2\lambda\tilde{F} + \lambda^2 \int_o^t \tilde{F}dT \tag{7-37}$$

where λ is the available control bandwidth. Differentiating (7-37) leads to

$$\dot{s} = \ddot{F}_E - \ddot{F}_d + 2\lambda\dot{\tilde{F}} + \lambda^2\tilde{F}$$

or equivalently, from (7-35) and (7-14), to

$$\dot{s} = -k_E m^{-1}(F + F_E + d) - \ddot{F}_d + 2\lambda\dot{\tilde{F}} + \lambda^2\tilde{F}$$

If $\dot{F}_E$ was precisely measured we would thus use

$$F = \hat{F} - \hat{m}\hat{k}_E^{-1}\,\overline{k}\,\,sat(s/\Phi)$$

where

$$\hat{F} = -F_E + \hat{m}\hat{k}_E^{-1}(-\ddot{F}_d + 2\lambda\dot{\tilde{F}} + \lambda^2\tilde{F}) \qquad (7-38)$$

and $\overline{k}$, Φ are obtained from gain margin β and gain k :

$$\beta = 2 \cdot \frac{3}{2} = 3 \qquad (7-39)$$

$$k = \hat{k}_E\hat{m}^{-1}\left[(\beta - 1)\,|\hat{F}| + \beta(D + \eta)\right] = \hat{k}_E\hat{m}^{-1}\left[2\,|\hat{F}| + 3(D + \eta)\right]\Bigg|_{(7-40)}$$

through the use of balance conditions (6-39)-(6-42). Again, note that "computed" control $\hat{F}$ is exactly of the form (7-19).

However, $\dot{F}_E$ is not measured, generally, but rather estimated according to (7-17) as

$$\dot{F}_E \approx -\hat{k}_E\,\dot{x}$$

Computed force $\hat{F}$ of (7-38) is then

$$\hat{F} = -F_E + \hat{m}\,\hat{k}_E^{-1}(-\ddot{F}_d - 2\lambda\dot{F}_d + \lambda^2\tilde{F}) - 2\lambda\hat{m}\dot{x}$$

so that control gain k of (7-40) becomes

$$k = \hat{k}_E \hat{m}^{-1} \left[2\,|\hat{F}| + 3(2\lambda\hat{m}\,|\dot{x}| + D + \eta) \right] \qquad (7\!-\!41)$$

The other effect of the uncertainty on $\dot{F}_E$ is that the value of the variable s itself is known only to within an accuracy of ϕ' :

$$\phi' = \hat{k}_E\,|\dot{x}| \ \geq\ |(\hat{k}_E - k_E)\dot{x}| = |\dot{\hat{F}}_E - \dot{F}_E|$$

From (7-35). Thus tracking precision in $\tilde{F}$ is only guaranteed to within

$$\epsilon' = 2(\phi + \phi')/\lambda \qquad (7\!-\!42)$$

Neglecting time-constants of order λ^{-1} , (7-42) can be written, from (7-39), (7-41) and the balance conditions, as

$$\epsilon' \approx (2\,\hat{k}_E \hat{m}^{-1}/\lambda)(\beta/\lambda) \left[2\,|\hat{F}_d| + 3(2\lambda\hat{m}\,|\dot{x}| + D + \eta) \right] + (2\,\hat{k}_E|\dot{x}|/\lambda)$$

where from (7-38)

$$\hat{F}_d = -F_E - \hat{m}\hat{k}_E^{-1}\ddot{F}_d$$

so that finally

$$\epsilon' \approx (12\,\hat{k}_E\hat{m}^{-1}\lambda^{-2})|F_E + \hat{m}\hat{k}_E^{-1}\ddot{F}_d| + (18\,\hat{k}_E\hat{m}^{-1}\lambda^{-2})(D + \eta) + 26\,\lambda^{-1}\hat{k}_E|\dot{x}|$$
$$(7\!-\!43)$$

Conversely (7-43) is the minimum value of $-F_d$ required for the contact between the manipulator and its environment to be maintained despite the presence of modeling uncertainty, or equivalently it is the minimum *bias force* to be applied in order to ensure that contact is maintained. This information is quite useful since applying excessive bias forces in order to ensure contact could damage the manipulator or the environment — of course it is in the multidimensional case that such a bound takes its real significance, as discussed later. Note that the first two terms of (7-43) vary as the inverse square of the control bandwidth λ, while the third term, due to the uncertainty on $\dot{F}_E$, only varies as the inverse of λ — this again reflects the sensitivity of the force control scheme to uncertainties on $\dot{F}_E$. Further, in (7-43) the extent D of disturbance d is amplified proportionally to $\hat{k}_E$ and inversely proportionally to $\hat{m}$, similarly to what was already noticed for the basic force control law of Section 7.3.1.

In the multidimensional case, the trajectory and force controllers of the hybrid control structure of Figure 7-8 can both be designed using the sliding control methodology. The resulting trajectory controller is analogous to that of Section 6.4 but is now based on end-effector dynamics (7-31) rather than joint-space dynamics (6-1), while the force controller is designed similarly using sliding surfaces of the form (7-37). The minimum bias force along each direction (analogous to the ϵ' of (7-43)) then becomes dependent on the uncertainty associated with the "computed" control structure (7-34), which leads to a trade-off between execution speed and minimum contact force.

7.4.2. Robustness in Dynamic Impedance Control

In Section 7.2.2, we saw that in principle impedance control exhibits natural robustness to parametric uncertainty. However, it is not always possible to apply control law (7-5) exactly because of uncertainty in the transmission mechanism, which plays a major role in compliant motion dynamics. Even for direct-drive arms, the effect of torque ripple (i.e. the nonlinearity of the relationship between motor current and torque generated) has to be properly accounted for, although one can easily show that it does not affect stability but only performance. Also, of course, Section 7.2.2 only considers the case of a *fixed* reference position $\mathbf{x}_d$. In this section, we briefly examine robustness issues in *dynamic* impedance control. Again, we first detail the one-dimensional case, and then generalize our results.

Consider the single degree-of-freedom system (7-13):

$$m\ddot{x} = F + F_E + d \qquad\qquad (7-44)$$

now required to exhibit a desired endpoint impedance behavior

$$k_D \dot{\tilde{x}} + k_P \tilde{x} = -F_E \qquad\qquad (7-45)$$

where $\tilde{x} = x - x_d$. Specification (7-45) can be interpreted as that of remaining on the sliding surface $s = 0$, with

$$s := \dot{\tilde{x}} + (k_P/k_D)\tilde{x} + F_E/k_D \qquad (7-46)$$

We can thus design a sliding controller that maintains the system as close as possible to the surface $s = 0$, given the available control bandwidth. The price to pay, however, is that *we now need an explicit measurement of interaction force F_E* , in order to compute the variable s. The sliding control law is of the form

$$F := -F_E + \hat{m}\,[\ddot{x}_d - (k_P/k_D)\dot{\tilde{x}} - (\dot{F}_E/k_D) - \overline{k}\ sat(s/\phi)\,] \qquad (7-47)$$

where the corresponding control discontinuity gain k accounts for the disturbance term d of (7-44), and for uncertainty on the "computed" term $\hat{m}\,[\ddot{x}_d - (k_P/k_D)\dot{\tilde{x}} - \dot{F}_E/k_D\,]$. Note that the control force F of (7-47) contains a term in $\dot{F}_E$, which is generally not measured directly, so that again an estimate of environment stiffness k_E is needed in order to use approximation (7-17) However, since the resulting uncertainty does not appear in the expression of s itself (as opposed to (7-37)) and further only contributes *additively* to the control input (as opposed to (7-38)), the precision of estimate k_E is not as critical here as it is in pure force control, and the effect of uncertainty on k_E can be directly added to the bounded disturbance term d. One can then easily show from the balance conditions that a minimum bias force $k_D(\beta^2 k_d^{max}/\lambda)$ is required in order to maintain contact in spite of the modeling uncertainties, where k_d^{max} is the maximum value of $k(x_d,\dot{x}_d)$, λ is the available control bandwidth, and β is the gain margin.

Note that the desired impedance (7-45) must itself be chosen "soft" enough not to excite the unmodelled dynamics, since F_E depends on the closed-loop dynamics of the global manipulator/environment system. Also, remark that when the manipulator is not in contact with the environment ($F_E = 0$), the sliding surface definition (7-46) reduces to that of a pure trajectory control problem.

The previous discussion can again be directly extended to the multidimensional case by rewriting manipulator dynamics in terms of end-effector coordinates. As in Section 7.4.1, the minimum bias force required along each direction is then dependent on the uncertainty associated with the computed control structure (7-34), which again yields a trade-off between task execution speed and minimum contact force.

7.5. Research Topics

Compliant motion control constitutes one of the most challenging and active areas of today's robotics research.

Robustness problems are particularly acute in compliant motion control, due to the strong effects of stiction, joint and link flexibility, and transmission backlash, as well as the necessity of some model of the environment. It may be feasible, however, to *adaptively* compensate for friction (e.g., Canudas *et al.*, 1986) or manipulator flexibility. Similarly, one may consider estimating the environment stiffness (as well as some its mass properties, if it is not fixed) on-line so as to improve the robustness of force control or the performance of dynamic impedance control.

Impedance control can in principle be extended in order to obtain a desired *apparent mass* or apparent inertia tensor (Hogan, 1985-a). However, this involves important stability issues, since one can easily verify that "positive" force feedback is needed in order to obtain an apparent mass smaller than the actual mass. A study of the robustness of impedance control in a linearized framework can be found in (Kazerooni, 1985), along with a discussion of target impedance specification.

Micropositioning had received a lot of interest recently (e.g. Asakawa *et al.*, 1982; Hollis, 1985) and consists in equipping the end-effector with a light, high-bandwidth, high-accuracy, actively controlled mechanical device. More generally, a challenging research topic consists in the design of *robot hands*, on which a comprehensive introduction can be found in (Mason and Salisbury, 1985). As in the case of flexible robots, a related and important problem is that of adequate sensing (e.g. Miller, 1986).

Fine motion planning is concerned with the effective sensor-based sequencing of subtasks in order to complete a particular assembly task (e.g. Lozano-Perez *et al.*, 1984; Mason and Salisbury, 1985). It is based on idealized friction models and the use of subtasks "backprojections" in a fashion similar to dynamic programming (Bellman, 1957). An important issue is again robustness to modeling approximations (Erdmann, 1985).

The effective integration of *machine vision* in robot control constitutes a whole research field in itself (Horn, 1986), as does *machine locomotion* (e.g. Raibert, 1986).

Finally, large research efforts are devoted to the study of specific industrial applications, such as grinding, deburring, polishing, or particular assembly tasks.

NOTES AND REFERENCES

More on the notion of natural and artificial constraints and its application to task-planning can be found in (Mason, 1981). The development of Section 7.2 is largely based on the results of (Salisbury, 1980; Hogan, 1981; Arimoto and Miyazaki, 1983; Koditschek, 1985; Slotine and Yoerger, 1986). A detailed discussion of the physical motivations and implications of impedance control can be found in Hogan (1985-a,b). The development of Section 7.3 is largely based on (Raibert and Craig, 1981; Khatib, 1985). Details on joint torque sensing can be found in e.g. (Asada and Lim, 1985), along with an application to the accurate torque control of a direct-drive arm. A comprehensive survey of compliant motion control schemes can be found in (Whitney, 1985).

PROBLEMS

CHAPTER 1

1.1 For the two-link manipulator of Figure 1-12, show that the endpoint position (x,y) can be expressed as a function of the joint angles θ_1 and θ_2 as follows:

$$x = l_1 \cos (\theta_1) + l_2 \cos (\theta_1 + \theta_2)$$
$$y = l_1 \sin (\theta_1) + l_2 \sin (\theta_1 + \theta_2)$$

Consider now the inverse problem of finding the joint angles required to obtain a desired endpoint position. Show that

$$\cos (\theta_2) = (x^2 + y^2 - l_1^{\,2} - l_2^{\,2})/(2\, l_1 l_2)$$

$$\theta_1 = \tan^{-1}(y/x) - \tan^{-1}\{ \,[l_2 \sin (\theta_2)]/[l_1 + l_2 \cos (\theta_2)] \, \}$$

and thus that there are two solutions ("elbow up" and "elbow down") depending on which quadrant θ_2 is chosen from.

CHAPTER 2

2.1 Use homogeneous transformations to derive the results of Problem 1.1.

2.2 A six degree-of-freedom robot manipulator PUMA 600[1] is shown in Figure P.2-1. Coordinate axes are assigned to each link according to the Denavit-Hartenberg convention.

---

[1] PUMA stands for "Programmable Universal Machine for Assembly". Originally part of a 1975 development project at General Motors Corp., PUMA's were later commercialized by Unimation, Inc.

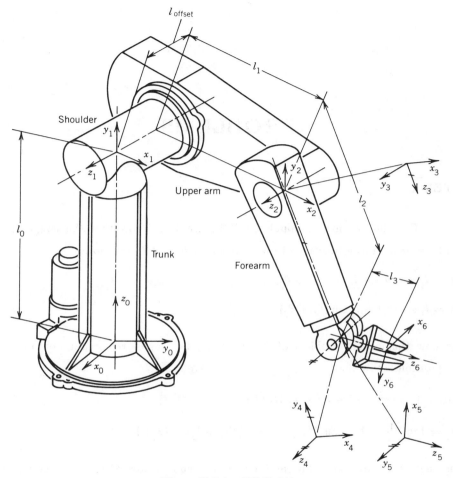

Figure P.2-1 : PUMA 600.

(1) Determine the link parameters α_i , a_i and d_i from the figure and set up the kinematic equation of the robot using 4×4 matrices.

(2) Solve the kinematic equation for joint displacements, given that the desired position and orientation of the endpoint are :

$$T = \begin{bmatrix} n_x & t_x & b_x & x_0 \\ n_y & t_y & b_y & y_0 \\ n_z & t_z & b_z & z_0 \\ 0 & 0 & 0 & 1 \end{bmatrix}$$

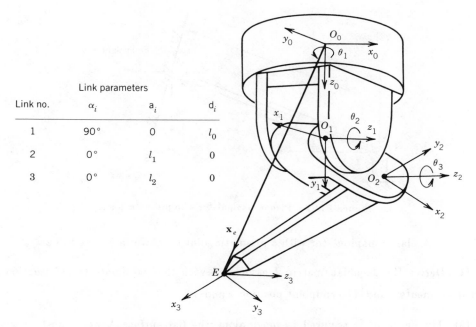

	Link parameters		
Link no.	α_i	a_i	d_i
1	90°	0	l_0
2	0°	l_1	0
3	0°	l_2	0

Figure P.3-1 : Three degree-of-freedom manipulator.

Use arc-tangent functions rather than arc-sines or arc-cosines.

(3) Assuming that each joint is allowed to rotate 360 degrees, discuss how many solutions exist for a given endpoint location.

CHAPTER 3

3.1   Consider the three-degree-of-freedom manipulator shown in Figure P.3-1. The link parameters are listed in the figure according to the Denavit-Hartenberg convention.

(1) Find the 3 × 3 Jacobian matrix associated with the transformation from joint displacements θ_1, θ_2, θ_3 to endpoint coordinates $\mathbf{x}_e = [x_e, y_e, z_e]^{\mathrm{T}}$.

(2) If each joint is allowed to rotate 360°, does the arm configuration get singular? If so, determine the endpoint positions corresponding singular configurations, and find for each position the directions along which the endpoint cannot move.

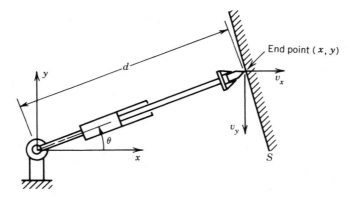

Figure P.3-2 : Planar manipulator with prismatic joint.

3.2   A planar manipulator with a prismatic joint is shown in Figure P.3-2.

(1) Derive the Jacobian matrix **J** associated with the coordinate transformation from joint displacements θ and d to endpoint position x and y .

(2) The endpoint is required to move along the flat surface Σ at a constant velocity (v_x, v_y). Compute the corresponding joint velocities and accelerations in terms of the joint displacements.

3.3   A planar manipulator with three revolute joints is shown below. Let θ and l_i be

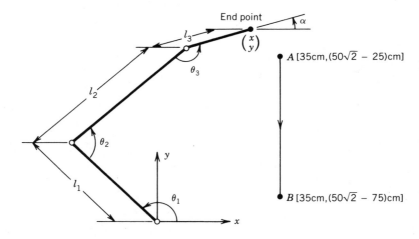

$l_1 = (50\sqrt{2} - 50)$cm, $l_2 = 25\sqrt{2}$ cm, $l_3 = 10$ cm

Figure P.3-3 : Planar manipulator with three revolute joints.

the angle of joint i and the length of link i, respectively, and x, y, and α be the endpoint position and orientation, as shown in the figure.

(1) Derive the manipulator Jacobian matrix.

(2) Compute endpoint velocities $\dot{x}$, $\dot{y}$, and $\dot{\alpha}$, given that joint angles and angular velocities are $\theta_1 = 120$ deg, $\dot{\theta}_1 = 5$ deg/s ; $\theta_2 = 60$ deg, $\dot{\theta}_2 = 10$ deg/s ; $\theta_3 = 90$ deg, $\dot{\theta}_3 = 20$ deg/s.

(3) The endpoint is required to move from point A to point B along the y axis at a constant speed of 10 cm/s. Assuming that link 3 is kept parallel to the x axis, compute the angular velocities of the three joints, and plot them as functions of time.

(4) Does the manipulator Jacobian become singular during the motion from point A to point B ? If so, determine the singular configuration and the direction along which the arm cannot move.

(5) If no condition is imposed on the orientation of link 3 during the motion starting at point A, determine the joint velocities at point A that minimize the squared norm $v^2 = \dot{\theta}_1^2 + \dot{\theta}_2^2 + \dot{\theta}_3^2$.

CHAPTER 4

4.1   The three wrist joints of a PUMA 600 are shown in Figure P.4-1. The robot is grinding a work surface, using a grinding tool grasped in its hand.

(1) The kinematic configuration of the wrist joints is defined in Table P.4-1, with reference to the coordinate frames shown in Figure P.4-1. The grinding tool is in contact with the work surface at point A, whose coordinates with reference to $O_3 - x_3 y_3 z_3$ are $x_3 = 10$ cm , $y_3 = 0$, and $z_3 = 5$ cm . Derive the 6 × 3 Jacobian matrix associated with the relationship between joint displacements and the position and orientation of the tool at point A.

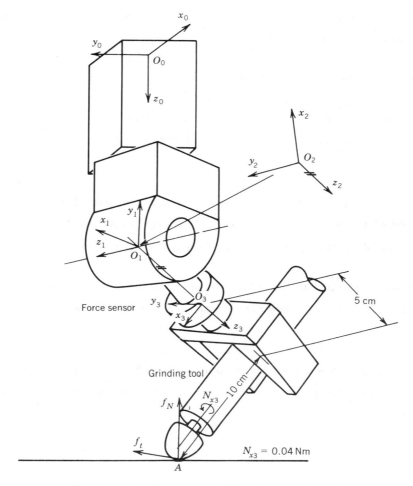

Figure P.4-1 : Wrist joints of PUMA with grinding tool.

(2) During the grinding operation, reaction forces and moments act on the tool tip A. Representing the forces and moments by a 6×1 vector $\mathbf{F}$, derive the equivalent joint torques. Also, compute the equivalent joint torques for the following case: The work surface is parallel to the x_0 and y_0 axes, and the normal force f_N and tangential force f_t along the z_0 and x_0 axes are -10N and -8N, respectively. The moment about the x_3 axis is 0.04 Nm in a right-hand sense. The joint angles are: $\theta_1 = 90^\circ$, $\theta_2 = 45^\circ$, and $\theta_3 = 0^\circ$.

(3) The robot has a force sensor attached to the origin of coordinate frame $O_3 - x_3 y_3 z_3$. The sensor measures three linear forces along the x_3, y_3, and z_3 axes, and three moments

Table P.4-1

link number	α_i	a_i	d_i
1	-90°	0	40 cm
2	+90°	0	0
3	0	0	10 cm

about these axes. Using these measured forces and moments, denoted by $f_{Mx}, f_{My}, f_{Mz}, N_{Mx}, N_{My}$ and N_{Mz}, respectively, find the forces and moments at the tool tip

$$\mathbf{F} = [f_{Tx} \ f_{Ty} \ f_{Tz} \ N_{Tx} \ N_{Ty} \ N_{Tz}]^T$$

with reference to $O_0 - x_0 y_0 z_0$.

4.2   Shown below is a two degree-of-freedom planar manipulator driven by hydraulic cylinders. The link lengths of the manipulator are denoted by $l_1 = O_0 O_i$ and $l_2 = O_i E$. The arm configuration is represented by the joint angles θ_1 and θ_2, as shown in the figure. The hydraulic cylinder HC_1 is connected between the two revolute joints, A and B, which are free to rotate and frictionless. As the length of the cylinder, $s_1 = AB$, varies, link 1 rotates about

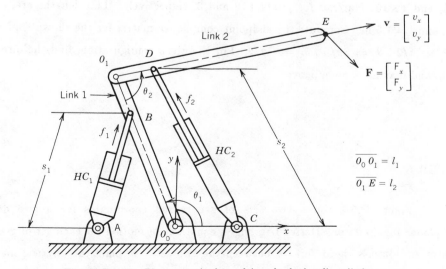

Figure P.4-2 : Planar manipulator driven by hydraulic cylinders.

point O_0 and joint angle θ_1 varies accordingly. The functional relationship between cylinder length s_1 and joint angle θ_1 is given by the differentiable function $\theta_1 = \theta_1(s_1)$. The hydraulic cylinder HC_2, on the other hand, is connected between revolute joint C on the base and another revolute joint D on link 2. The joint angle θ_2 varies with the cylinder length $s_2 = CD$. It also varies with the cylinder length s_1, and therefore is a function of both s_1 and s_2: $\theta_2 = \theta_2(s_1, s_2)$. The function θ_2 is differentiable with respect to both s_1 and s_2.

(1) When the lengths of cylinders HC_1 and HC_2 vary at speeds $\dot{s}_1$ and $\dot{s}_2$, respectively, find the endpoint velocity $\mathbf{v} = [v_x, v_y]^T$ with reference to the base coordinate frame O_0-xy. Obtain velocity $\mathbf{v}$ using the functions $\theta_1(s_1)$ and $\theta_2(s_1, s_2)$, and their derivatives at a given instant when the cylinder lengths are s_1 and s_2.

(2) Let f_1 and f_2 be the forces exerted by the cylinders HC_1 and HC_2, respectively. Each force acts in the longitudinal direction of the cylinder, and is defined to be positive when the cylinder expands. Assuming that all the joints are frictionless, find the cylinder forces f_1 and f_2 equivalent to a given endpoint force $\mathbf{F} = [F_x, F_y]^T$, defined with reference to O_0-xy, i.e., the cylinder forces that balance the negative endpoint force $-\mathbf{F}$ when gravity forces and inertial forces are ignored.

4.3   Consider again the 3 degree-of-freedom manipulator shown in Figure P.3-1. The joint servo stiffness are measured at individual joints. They are 4×10^5 Nm/rad, 2×10^5 Nm/rad, and 1×10^5 Nm/rad for joints 1, 2 and 3, respectively. Link lengths are: $l_0 = 1$ m, $l_1 = 1$ m, and $l_2 = 1.5$ m. Compute the endpoint compliance matrix for the shown configuration, where $\theta_1 = \pi/2$, $\theta_2 = 3\pi/4$, $\theta_3 = -\pi/2$. For the given configuration, find the directions of minimum and maximum compliance.

CHAPTER 5

5.1   Figure P.5-1 shows two sets of generalized coordinates for a two degree-of-freedom planar manipulator. Derive Lagrange's equations of motion for each set of generalized coordinates and discuss the difference between the two sets. Explain the physical meaning of generalized forces in each case.

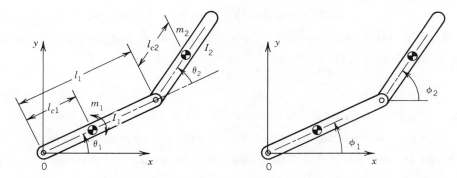

Figure P.5-1 : Two sets of generalized coordinates for a two degree-of-freedom planar manipulator.

5.2   Consider the two degree-of-freedom planar manipulator shown below. The length of link i is denoted by l_i and its mass by m_i. The distance between joint i and the centroid of link i is denoted by l_{ci} and the centroidal moment of inertia is given by I_i. When the manipulator grasps an unknown mass particle M at the tip of link 2, the mass properties of link 2 change from known values m_2, l_{c2}, I_2 to unknowns m_2^*, l_{c2}^*, I_2^*, where $m_2^* = m_2 + M$. It is required to identify the unknown mass properties through experiments.

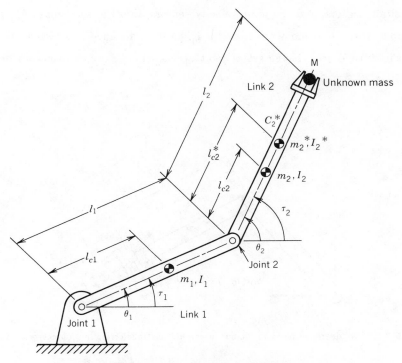

Figure P.5-2 : Two degree-of-freedom planar manipulator.

(1) The unknown mass M is modeled as a point mass, and the centroids are assumed to be located on the center line of each link. Derive the distance l_{c2}^{*} and the centroidal moment of inertia I_2^{*} as functions of unknown mass.

(2) At a given instant, the generalized coordinates are θ_1 and θ_2, the actuators exert torques τ_1 and τ_2, and the manipulator with the unknown mass moves at angular velocities $\dot{\theta}_1$, $\dot{\theta}_2$ and angular accelerations $\ddot{\theta}_1$, $\ddot{\theta}_2$. Determine the unknown mass M at the tip of link 2 from this set of data.

5.3   Figure P.5-3 shows the manipulator with a gripper transferring an object whose mass properties are unknown. In order to identify the mass properties, a wrist sensor detecting linear forces, f_u and f_v, and moment N_w, is attached to the arm tip as shown in the figure. Joint angles, joint velocities and accelerations, as well as forces and moment at the wrist sensor, are measured while moving the object. From the data measured by the wrist sensor, the joint position sensors, and the joint velocity sensors, find the total mass M_0 of the object and the gripper, their total moment of inertia I_0, and the distance l_0 between their total mass center and the wrist sensor. Is the set of joint accelerations data necessary to determine the mass properties?

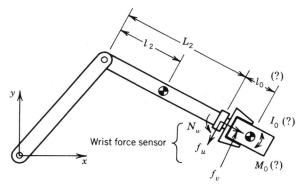

Figure P.5-3 : Two degree-of-freedom planar manipulator transferring an object with unknown mass properties.

Table P.5-1

	link 1	link 2
$l_i \ [m]$	1	1
$l_{ci} \ [m]$	0.5	0.5
$m_i \ [kg]$	20	10
$I_i \ [kgm^2]$	0.8	0.2
$\theta_i \ [deg]$	30	30
$\dot{\theta}_i \ [deg/s]$	15	45
$\ddot{\theta}_i \ [deg/s^2]$	-30	30

5.4   Table P.5-1 shows the link dimensions and mass properties of the two degree-of-freedom manipulator of Figure P.5-1. Using the recursive Newton-Euler formulation of Luh, Walker and Paul, compute the joint torques required to move the arm with the specified joint velocities and accelerations at the configuration given in the table.

5.5   Figure P.5-4 shows the same manipulator as in Figure P.5-1. The endpoint of the manipulator is in contact with the smooth surface and applying a normal force f_N to the surface, while moving in the tangential direction at a constant speed v_t. Compute the required joint torques, τ_1 and τ_2, in the case shown in the figure.

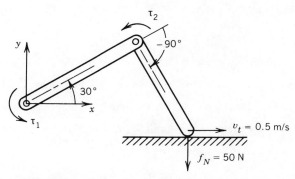

Figure P.5-4 : Two degree-of-freedom planar manipulator.

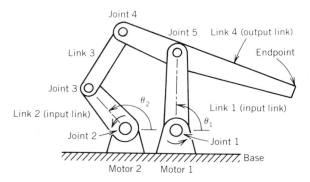

Figure P.5-5 : Planar manipulator with five-bar-linkage mechanism.

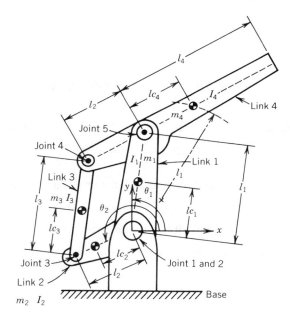

Figure P.5-6 : Planar manipulator with parallelogram mechanism.

5.6   Figure P.5-5 shows a manipulator arm consisting of a closed-loop planar five-bar-link mechanism. Two actuators are fixed to the base link and drive the two input links, link 1 and link 2. Link 4 is the output link, where an end-effector is attached. When the distance between joint 1 and 2 is set to zero, and links 1 through 4 form a parallelogram, the arm mechanism reduces to the one shown in Figure P.5-6.

(1) The angles θ_1 and θ_2 in the figure can be used as a complete and independent set of

generalized coordinates. Find the 2×2 manipulator inertia tensor in terms of θ_1 and θ_2, and show that the moment of inertia seen by each actuator is invariant for all the arm configurations.

(2) Show that, if the mass properties and link lengths of the mechanism satisfy the condition

$$m_4 \, l_1 \, l_{c4} = m_3 \, l_2 \, l_{c3}$$

then the two input joints do not have dynamic interactions and that no Coriolis and centrifugal torques are generated. Hence, the manipulator inertia tensor becomes constant and diagonal, so that the arm dynamics exhibits no coupling between the two actuators and no nonlinear effects except gravity.

CHAPTER 6

6.1 How would Proof 6.2 be affected if the desired trajectory $\mathbf{q}_d$ was time-varying?

6.2 Show that (6-10) indeed guarantees asymptotic stability in the presence of friction and gravity, provided that the gains k_{jI}, k_{jP}, and k_{jD} are within appropriate bounds, and in particular that k_{jI} is *small* enough (yet strictly positive).

Note: This exercise is fairly involved mathematically. You may wish to use a Lyapunov function candidate of the form

$$V = \frac{1}{2} [\dot{\mathbf{q}}^T \mathbf{H} \dot{\mathbf{q}} + \bar{\mathbf{q}}^T \mathbf{M} \bar{\mathbf{q}}]$$

where

$$\bar{\mathbf{q}} := \left(\begin{matrix} \tilde{\mathbf{q}} \\ \int_0^t \tilde{\mathbf{q}}(T) dT \end{matrix} \right)$$

and $\mathbf{M}$ is symmetric positive definite.

6.3   Consider the system (6-15), and assume now that $x^{(n)}$ is also available for measurement. This value can be used to compute $\dot{s}$ and include an additional term of the form $-a\dot{s}$ (with $a > 0$) in the control action u.

(1) Write the corresponding expression of $\dot{s}$, and assume for simplicity that the gain margin β is constant. Show that, for a given control bandwidth λ , the effect of the additional term in u is to reduce the maximum value of s by a factor $(1 + a\beta^{-1})$. Show that this implies that tracking error $\tilde{x}$ can in principle be made arbitrarily small by simply increasing the value of a.

(2) What are the problems of this approach? In particular, assume that there is an $I\%$ uncertainty on the value of $x^{(n)}$. How small must I be to make the approach worthwile?

6.4   Show that the balance conditions (6-39)-(6-41) can be digitally implemented in the general case as

$$a_1 = \lambda\phi/\beta_d - k_d$$

If $(a_1 \geq 0)$, $a_2 = \exp(-\lambda T/\beta_d^2)$; else $a_2 = \exp(-\lambda T)$.

$$\phi = a_2\phi + (1 - a_2)(\beta_d k_d/\lambda)$$

$$\overline{k} = k + a_1$$

where $k = k(\mathbf{X})$, $k_d = k(\mathbf{X}_d)$, $\overline{k} = \overline{k}(\mathbf{X})$, T is the sampling period, and (following usual computer practice) the notation $=$ assigns to the variable on the left the new value computed on the right.

6.5   Consider the two-link manipulator of Example 6-6, but now with an uncertainty consisting of an unknown load at the tip of the arm. The mass of the load is bounded by 1 . The structural resonant mode of lowest frequency is at 10 Hz. Compare in simulations the performance of a local P.I.D. scheme, a pure computed torque, and a sliding controller, in the following cases:

(1) $\theta_{d1} = -\pi/3 + \pi/3 \,[1 - \cos{(\pi t/T)}]$; $\theta_{d2} = 2\pi/3 - \pi/3 \,[1 - \cos{(\pi t/T)}]$

for $0 \leq t \leq T$, in the three cases: $T = 1$; $T = 0.5$; and $T = 2$. What is the minimum sampling rate required to implement your controller?

     (2) The desired trajectory is a straight line from $(x,y) = (1,0)$ to $(x,y) = (0,1.5)$ to be completed "at constant speed" within two seconds. The manipulator is initially at rest at $(x,y) = (1,0)$ in an "elbow down" configuration.

Note: When the time derivatives of the reference trajectory are not explicitly available, you may wish to generate $\mathbf{q}_d$ by smoothing the reference trajectory through a second-order lowpass filter of bandwidth λ , which then also provides $\dot{\mathbf{q}}_d$ and $\ddot{\mathbf{q}}_d$ explicitly.

CHAPTER 7

7.1   For the two-link manipulator of Figure 6.7, choose a joint stiffness matrix that places the compliance center at the tip of the arm, and such that the manipulator be stiff when pushed towards its shoulder (i.e., towards the origin) and compliant in the orthogonal direction.

7.2   How would the development of Section 7.2.2.1 be affected if desired trajectory $\mathbf{x}_d$ was time-varying ?

7.3   What are the effects of viscous and Coulomb friction on the development of Section 7.2.2.1 ?

7.4   What are the effects of viscous friction (at the manipulator joints) and Coulomb frition (both at the manipulator joints and at the point of contact with the environment) on the development of Section 7.2.2.2 ?

7.5   For the two-link articulated manipulator of Figure 6-6, design a hybrid controller so as to follow the environment surface:

$$(\Sigma): \quad x = \cos{(4\xi)} \quad ; \quad y = \sin{(\xi)}$$

according to $\xi = t$, while maintaining a constant contact force F_d . The environment surface

(Σ) is assumed to be frictionless, and the manipulator dynamic model is assumed to be precisely known. Draw a detailed block diagram of the complete controller.

7.6   Compute the control discontinuity gain k corresponding to equation (7-46) explicitly. Compare the performance of (7-46) with that of a pure impedance control scheme in a simple simulation. Assuming that the desired damping is constant, examine the variations in comparative performance as a function of k_P.

BIBLIOGRAPHY

Abel, J.M., Holzmann, W. and McCarthy, J.M., On Grasping Planar Objects with Two Articulated Fingers, *I.E.E.E. Int. Conf. Robotics and Automation*, St. Louis (1985).

Ahuja, N., Chien, R. T., Bridwell, N., Interference Detection and Collision Avoidance Among Three Dimensional Objects, *1st Ann. Nat. Conf. Artificial Intelligence, Stanford University* (1980).

Albus, J. S., A New Approach to Manipulator Control: The Cerebellar Model Articulation Controller (CMAC), *A.S.M.E. J. Dynamic Systems Measurement, Control*, **97** (1975).

Albus, J. S., Data Storage In The Cerebellar Model Articulation Controller (CMAC), *A.S.M.E. J. Dynamic Systems, Measurement, Control*, **97** (1975).

Ambler, A. P. and Popplestone, R. J., Inferring the Positions of Bodies From Specified Spatial Relationships, *Artificial Intelligence*, **6**, 2 (1975).

Ambler, A. P., Popplestone, R. J. and Kempf, K. G., An Experiment In the Off-Line Programming of Robots, *12th Int. Symp. Industrial Robots*, Paris (1982).

Andreev, G. Y., Assembling Cyclindrical Press Fit Joints, *Russian Engineering Journal*, **52**, 7 (1972).

Andreev, G. Y. and Laktionev, N. M., Problems In Assembly for Large parts, *Russian Engineering Journal*, **46**, 1 (1966).

Andreev, G. Y. and Laktionev, N. M., Contact Stress During Automatic Assembly, *Russian Engineering Journal*, **49**, 11 (1969).

Anex, R. P., and Hubbard, M., Modeling and Adaptive Control of a Mechanical Manipulator, *A.S.M.E. J. Dyn. Syst. Meas. Contr.*, **106**, 3 (1984).

Apple, H. P. and Lawrence, P. D. A Seven Degree of Freedom Digital Incremental Electric Orthotic Arm, *8th Ann. Int. Conf. Medical and Biological Engineering*, Chicago (1969).

Arai, T. and Kinoshita, N., The Part Mating Forces That Arise When Using a Worktable With Compliance, *Assembly Automation*, **6**, 4 (1981).

Arimoto, S., Learning Control Theory for Dynamic Systems - A Survey, *I.E.E.E. Conf. Decision and Control*, Fort Lauderdale (1985).

Arimoto, S. and F. Miyazaki, Stability and Robustness of P.I.D. Feedback Control for Robot Manipulators of Sensory Capability, *1st Int. Symp. Robotics Res.* (1983).

Aristova, M. V., Ignatiev, M. B. and Prokhorov, V. M., Algorithmic System for Robot's Motion Simulation, *Symp. Theory and Practice of Robots and Manipulators*, Warsaw, Poland (1976).

Armstrong, W. W., Recursive Solution to The Equations of Motion of an N-lind Manipulator, *5th World Congress On Theory of Machines and Mechanism*, Montreal (1979).

Asada, H., Studies In Prehension and Handling By Robot Hands With Elastic Fingers, *Doctoral Dissertation, Kyoto University* (1979).

Asada, H., A Geometrical Representation of Manipulator Dynamics and its Application to Arm Design, *A.S.M.E. J. of Dynamic Systems, Measurement, and Control*, **105**, 3 (1983).

Asada, H., and By, A. B., Kinematic Analysis and Design for Automatic Workpart Fixturing in Flexible Assembly, *2nd Int. Sym. of Robotics Research*, Kyoto (1984).

Asada, H., and By, A. B., Kinematic Analysis of Workpart Fixturing for Flexible Assembly with Automatically Reconfigurable Fixtures, *I.E.E.E. J. of Robotics and Automation*, **1**, 2 (1985).

Asada, H. and Cro-Granito, J.A., Kinematic and Static Charaterization of Wrist Joints and Their Optimal Design, *I.E.E.E. Int. Conf. on Robotics and Automation*, St. Louis (1985).

Asada, H., and Goldfine, N., Optimal Compliance Design for Grinding Robot Tool Holders, *I.E.E.E. Int. Conf. On Robotics and Automation*, St. Louis (1985).

Asada, H., Kanade, T., Design of Direct-Drive Arms, *A.S.M.E. Journal of Vibration, Stress, and Reliability In Design*, **105**, 3, July (1983).

Asada, H., Kanade, T., Takeyama, I., Control of A Direct-Drive Arm, *A.S.M.E. Journal of Dynamic Systems, Measurement, and Control* **105**, 3 (1983).

Asada, H. and Lim S.K., Design of Joint Torque Sensors and Torque Feedback Control for Direct Drive Arms, *A.S.M.E. Winter Annual Meeting*, Miami (1985).

Asada, H., and Ro, I. H., A Linkage Design for Direct-Drive Arms, *A.S.M.E. J.. of Mech., Trans., Auto. in Design*, **107**, 4 (1985).

Asada, H., and Youcef-Toumi, K., Analysis and Design of Semi-Direct Drive Robot Arms, *American Control Conf.*, San Francisco (1983).

Asada, H. and Youcef-Toumi, K., Analysis and Design of A Direct-Drive Arm With A Five-Bar Link Parallel Drive Mechanism, *A.S.M.E. J. Dyn. Syst. Meas. Contr.*, **106**, 3 (1984).

Asakawa, K., Akiya F., and Tabata F., A Variable Compliance Device and Its Application for Automatic Assembly, *Fujitsu Laboratories*, Kawasaki, Japan (1982).

Åström, K.J., Theory and Applications of Adaptive Control: A Survey, *Automatica 19 (5)* (1983).

Åström, K. J. and Wittenmark B., *Computer Controlled Systems: Theory and Design*, Prentice-Hall (1984).

Atkeson, C. G., An, C. M. and Hollerbach, J. M., Estimation of Inertial Parameters of Manipulator Loads and Links, *Int. Symp. Robotics Res.*, Gouvieux (1985).

Baker, B.S., Fortune, S., and Gross, E., Stable Prehension with a Multi-Fingered Hand, *I.E.E.E. Int. Conf. on Robotics and Automation*, St. Louis (1985).

Bejczy, A. K., Robot Arm Dynamics and Control, *Jet Propulsion Laboratory, California Institute of Technology, TM 33-69*, (1974).

Bejczy, A. K., Dynamic Models and Control Equations for Manipulators, *Jet Propulsion Laboratory, California Institute of Technology*, **715**, 19 (1979).

Bejczy, A. K. and Lee, S., Robot Arm Dynamic Model Reduction for Control, *I.E.E.E. Conf. Decision and Control*, San Antonio (1983).

Bejczy, A. K. and Paul, R. P., Simplified Robot Arm Dynamics for Control, *I.E.E.E. Decision and Control Conf.*, San Diego (1981).

Bellman, R.E., Dynamic Programming, *Princeton University Press*, Princeton, NJ (1957).

Ben Israel, A. and Greville, T. N., Generalized Inverses: Theory and Applications, *John Wiley and Sons*, New York, (1974)

Binford, T. O., Visual Perception By Computer, *I.E.E.E. Conf. Systems and Control*, (1971).

Binford, T., et al., Exploratory Studies of Computer Integrated Assembly Systems: NSF Progress Reports, *Artificial Intelligence Laboratory, Stanford University*, September 1974, November 1975, July 1976.

Bobrow, J. E., Optimal Control of Manipulators, *Doctoral Dissertation, University of California, Los Angeles* (1982).

Bobrow, J. E., Dubowsky, S., and Gibson, J. S., On the Optimal Control of Robotic

238   **Bibliography**

Manipulators with Actuator Constraints, *American Control Conf.*, San Francisco (1983).

Bobrow, J. E., Dubowsky, S., and Gibson, J. S., Time-Optimal Control of Robotic Manipulators Along Specified Paths, *Int. J. Robotics Research*, **4**, 3 (1985).

Bondarev, A.G., Bondarev, S.A., Kostylyova, N.Y., and Utkin, V.I., Sliding Modes in Systems with Asymptotic State Observers, *Autom. Remote Contr.*,**6** (1985) (in Russian).

Book, W. J., Recursive Lagrangian Dynamics of Flexible Manipulator arms Via Transformation Matrices, *I.F.A.C. Symp. CAD Multivariable Technological Sys.*, W. Lafayette, Indiana, **5**, 17 (1983).

Book W. J., Recursive Lagrangian Dynamics of Flexible Manipulator Arms, *2nd Int. Sym. of Robotics Research*, Kyoto (1984).

Book, W. J., Maizza-Neto, O. and Whitney, D. E., Feedback Control of Two-Beam Two-Joint Systems With Distributed Flexibility, *J. Dynamic Systems, Measurement and Control* (1976).

Bottema, O. and Roth, B., Theoretical Kinematics,*North Holland, Amsterdam* (1979).

Brady, M. T., Hollerbach, M., Johnson, T. L., Lozano-Perez, T. and Mason, M. T., Robot Motion, *M.I.T. Press* (1982).

Brooks, R. A., Solving The Find-path Problem By Representing Free space As Generalized Cones, *Artificial Intelligence Laboratory, Massachusetts Institute of Technology* (1982a).

Brooks, R. A., Symbolic Error Analysis and Robot Planning, *Int. J. Robotics Research*, 1, 4 (1982b).

Brooks R. A. and Lozano-Perez, T., A Subdivision Algorithm In Configuration Space for Findpath With Rotation, *I.E.E.E. Trans. Systems, Man, and Cybernetics*, **15**, 2 (1985).

Brou, P., Implementation of High-Level Commands for Robots, *M.S. Thesis, Department of Electrical Engineering and Computer Science, Massachusetts Institute of Technology* (1980).

Burstall, R. M., and Collins, J. S. and Popplestone, R. J., Programming In POP-2, *Edinburgh University Press*, Edinburgh (1971).

Campbell, C. E. and Luh, J. Y. S., A Preliminary Study On Path Planning of Collision Avoidance for Mechanical Manipulators, *School of Electrical Engineering, Purdue University*, **80**, 48 (1980).

Cannon, R. H., and Schmitz, E., Initial Experiments On The End-Point Control of A Flexible One-Link Robot, *Int. J. Robotics Research*, **3**, 3 (1984).

Canudas, C., Åström, K. J. and Braun, K., Adaptive Friction Compensation in D.C.

Motor Drives, *I.E.E.E. Int. Conf. Robotics and Automation*, San Francisco (1986).

Chua, L. O., and Lin G. N., Nonlinear Programming Without Computation, *I.E.E.E. Trans. Circuits and Systems*, **31**, 2(1984).

Craig, J., Adaptive Control of Manipulators through Repeated Trials, *American Control Conf.*, San Diego (1984).

Cwiakala, M. and Lee, T. W., Generation and Evaluation of a Manipulator Workspace Based on Optimum Path Search, *A.S.M.E. J. Mech., Trans., & Auto. in Design*, **107**, 2 (1985).

Datseris, P. and Palm, W., Principles on the Development of Mechanical Hands Which Can Manipulate Objects by Means of Active Control, *A.S.M.E. J. Mech., Trans., & Auto. in Design*, **107**, 2 (1985).

De Fazio, T. L., Displacement State Monitoring for The Remote Center Compliance (RCC) — Realizations and Applications,*10th Int. Symp. Industrial Robots*, Milan (1980).

De Kleer, J., Qualitative and Quantitative Knowledge In Classical Mechanics, *Artificial Intelligence Laboratory, Massachusetts Institute of Technology*, 352 (1975).

Denavit, J. and Hartenberg, R. S., A Kinematic Notation for Lower Pair Mechanisms Bases On Matrices, *J. Applied Mechanics 22* (1955).

Desa, S. and Roth, B., Synthesis of Control Systems for Manipulators Using Multivariable Robust Servomechanism Theory, *Int. J. Robotics Research*, **4**, 3 (1985).

Dillon, S. R., Computer Assisted Equation Generation In Linkage Dynamics, *Doctoral Dissertation, Ohio State University* (1973).

Drake, S. H., Using Compliance In Lieu of Sensory Feedback for Automatic Assembly, *Doctoral Dissertation, Department of Mechanical Engineering, Massachusetts Institute of Technology* (1977).

Dubowsky, S. and DesForges, D. T., The Application of Model-referenced Adaptive Control to Robotic Manipulators, *J. Dynamic Systems, Measurement, Control*, **101** (1979).

Dubowsky, S. and Sunada, W. H., On The Dynamic Analysis and Behavior of Industrial Robotic Manipulators With Elastic Members, *A.S.M.E. Journal of Mechanisms, Transmissions, and Automation In Design*, **105**, 1 (1983).

Duffy, J., Displacement Analysis of Spatial Seven-line 5R-2P Mechanisms, *J. Applied Mechanics* (1976).

Duffy, J. and Rooney, J., Displacement Analysis of Spatial Six-line, 5R-C Mechanisms, *J. Applied Mechanics*, **41** (1974).

Engelberger, J. F., Four Million Hours of Robot Field Experience, *4th Int. Symp. Industrial Robots*, Tokyo (1974).

Engelberger, J. F., Robotics In Pratice: Management and Applications of Industrial Robots,*Amacon* (1980).

Erdmann, M.A., Using Backprojections for Fine Motion Planning with Uncertainty, *I.E.E.E. Int. Conf. Robotics and Automation*, St. Louis (1985).

Ernst, H. A., MH-1: A Computer-Operated Mechanical Hand, *Sc. D. Thesis, Massachsuetts Institute of Technology* (1961).

Eun, T., Cho, Y. J. and Cho, H. S., Stability and Positioning Accuracy of A Pneumatic On-off Servomechanism, *American Control Conf.*, Alexandria, VA (1982).

Fahlman, S. E., A Planning System for Robot Construction Tasks, *Artificial Intelligence*, **5**, 1 (1974).

Faux, I. D. and Pratt, M. J., Computational Geometry for Design and Manufacture, *Ellis Horwood Press, Chichester* (1979).

Fearing, R. S. and Hollerbach, J. M., Basic Solid Mechanics for Tactile Sensing, *Int. J. Robotics Research*, **4**, 3 (1985).

Featherstone, R., The Calculation of Robot Dynamics Using Articulated-Body Inertias, *Int. J. Robotics Research*, **2**, 1, (1983-a).

Featherstone, R., Position and Velocity Transformations Between Robot End Effector Coordinates and Joint Angles, *Int. J. Robotics Research*, **2**, 1 (1983-b).

Feynman, R.P., Leighton R.B. and Sands M., The Feynman Lectures On Physics, *Addison-Wesley* (1963).

Feynman, R.P., Quantum Mechanical Computers, *Optics News* (1985).

Filippov, A. F., *Am. Math. Soc. Trans.*, **62**, 199 (1960)

Finkel, R. A., Constructing and Debugging Manipulator Programs, *Artificial Intelligence Laboratory*, Stanford University, 284 (1976).

Finkel, R., Taylor, R., Bolles, R., Paul, R. and Feldman, J., AL, A Programming System for Automation, *Artificial Intelligence Laboratory*, Stanford Univeristy, 177 (1974).

Freudentein, F., Longman, R. W., and Chen, C. K., Kinematic Analysis of Robotic Bevel-Gear Trains, *A.S.M.E. J.. of Mech., Trans., Auto. in Design*, **106**, 3 (1984).

Freund, E., Decoupling and Pole Assignment In Nonlinear Systems, *Electronics Letter*, **9**, 16 (1973).

Freund, E., Fast Nonlinear Control With Arbitrary Pole-placement for Industrial Robots and Manipulators, *Int. J. Robotics Research* **1**, 1 (1982).

Freund, E. and Syrbe, M., Control of Industrial Robots by Microprocessors, *IRIA Conf. New Trends In Systems Analysis*, Rocquencourt, France (1976).

Gavrilovic, M. M. and Maric, M. R., An Approach to the Organization of the Artificial Arm Control, *3rd Int. Symp. External Control of Human Extremities*, Dubrovnic (1969).

Gilbert, E. G. and Ha, I. J., An Approach to Nonlinear Feedback Control With Applications to Robotics, *I.E.E.E. Conf. Decision and Control*, San Antonio (1983).

Gini, G., Gini, M., Gini, R., and Giuse, D., Introducing Software Systems In Industrial Robots, *9th Int. Symp. Industrial Robots*, Washington D.C. (1979).

Giralt, G., Sobek R.,and Chatila R., A Multilevel Planning and Navigation System for A Mobile Robot, *6th Int. Joint Conf. Artificial Intelligence*, Tokyo, Japan (1979).

Golden, B., Shortest Path Algorithms: A Comparison, *Operations Research*, **24**, 6 (1976).

Goldstein, H. Classical Mechanics, Addison-Wesley (1981).

Golla, D. F., Linear State-feedback Control of Rigid-link Manipulators, *University of Toronto, Insitute for Aerospace Studies* (1979).

Golla, D. F., Garg, S. C. and Hughes, P. C., Linear State-feedback Control of Manipulators, *Mech. Machine Theory 16* (1981).

Goto, T., Inoyoma, T. and Takeyasu, K., Precise Insert Operation By Tactile Controlled Robot, *2nd Conf. Industrial Robot Technology* (1974).

Goto, T., Takeyasu, K. and Inoyama, T., Control Algorithm for Precision Insert Operation Robots, *I.E.E.E. Trans., Systems Man, Cybernetics*, **10**, 1 (1980).

Groome, R. C., Force Feedback Steering of A Teleoperator System, *M.S. Thesis, Department of Aeronautics and Astronautics*, Massachusetts Institute of Technology (1972).

Grossman, D. D. and Taylor, R. H.,Interactive Generation of Object Models With a Manipulator, *I.E.E.E. Trans. Systems, Man, Cybernetics* **8**, 9 (1978).

Gupta, K. C., and Roth, B., Design Considerations for Manipulator Workspace, *A.S.M.E. J.. Mechanical Design*, **104**, 4 (1982).

Gusev, A. S., Automatic Assembly of Cylindrically Shaped Parts, *Russian Engineering Journal*, **49**, 11 (1969).

Ha, I.J. and Gilbert, E.G., Robust Tracking in Nonlinear Systems and its Applications to Robotics, *I.E.E.E. Conf. Decision and Control*, Fort Lauderdale (1985).

Hanafusa, H. and Asada, H., Mechanics of Gripping Form by Artificial Fingers, *Trans. Society of Instrument andf Control Engineers*, **12**, 5 (1976).

Hanafusa, H., and Asada, H., A Robotic Hand With Elastic Fingers and its Application to Assembly Process, *I.F.A.C. Symp. Information and Control Problems in Manufacturing Technology*, Tokyo (1977-a).

Hanafusa, H. and Asada, H., Stable Prehension of Objects by the Robot Hand with Elastic Fingers, *7th Int. Symp. Industrial Robots*, Tokyo (1977-b).

Hanafusa, H., Yoshikawa, T., and Nakamura, Y., Analysis and Control of Articulated Robot Arms with Redundancy, *Prep. 8th IFAC World Congress* (1981).

Harmon, L. D., Automated Tactile Sensing, *Int. J. Robotics Research*, **1**, 2 (1982).

Harris, T. A., Roller Bearing Analysis, *J. Wiley & Sons*, New York (1966).

Hart, P., Nilsson, N. and Raphael, B., A Formal Basis for the Heuristic Determination of Minimum Cost Paths, *I.E.E.E. Trans. System Science and Cybernetics*, **4**, 2 (1968).

Hartenberg, R. S. and Denavit, J., Kinematic Synthesis of Linkages, *McGraw-Hill*, New York (1964).

Hasegawa, T., A New Approach to Teaching Object Descriptions for A Manipulation Environment, *12th Int. Symp. Industrial Robots*, Paris, France (1982).

Hastings, G. G., and Book, W. J., Experiments in Optimal Control of a Flexible Arm, *American Control Conf.*, Boston (1985).

Heer, E. (Editor), Remotely Manned Systems — Exploration and Operation in Space, *1st National Conf. Remotely Manned Systems*, California Institute of Technology (1971).

Heginbotham, W. B., Page, C. J. and Paugh, A., Robot Research at the University of Nottingham, *4th Int. Symp. Industrial Robots*, Tokyo (1974).

Heginbotham, W. B., Dooner, M. and Case, K., Robot Application Simulation, *Industrial Robot* (1979).

Hemami, H. and Jaswa, V. C., On a Three-link Model of the Dynamics of Standing Up and Sitting Down, *I.E.E.E. Trans Systems, Man and Cybernetics*, **8**, 2 (1978).

Hemami, H., Jaswa, V. C. and McGhee, R. B., Some Alternative Formulations of Manipulator Dynamics for Computer Simulation Studies, *Allerton Conf. Circuit and System Theory*, University of Illinois (1975).

Hilbert, D., and Cohn-Vossen, S., Geometry and The Imagination, *Chelsea* (1952).

Hillis, D., A High-Resolution Image Touch Sensor, *Int. J. Robotics Research*, **1**, 2 (1982).

Ho, J. Y., Direct Path Method for Flexible Multibody Spacecraft Dynamics, *A.I.A.A. J. Spacecraft and Rockets*, **14**, 2 (1977).

Hogan, N., Control of Mechanical Impedance of Prosthetic Joints, *Joint Automatic Control Conf.*, San Francisco (1980).

Hogan, N., Impedance Control of a Robotic Manipulator, *Winter Ann. Meeting of the A.S.M.E.*, Washington, D.C. (1981).

Hogan, N., Impedance Control: An Approach to Manipulation, *J. Dyn. Syst., Meas., and Control*, **107** (1985a).

Hogan, N., Control Strategies for Computer Movements Derived from Physical Systems Theory, *Int. Symp. Synergetics*, Bavaria (1985b).

Hogan, N. and Mann, R. M., Myoelectric Signal Processing: Optimal Estimation Applied to Electroyography, Parts I and II, *I.E.E.E. Trans. Biomedical Engineering*, **27**, 7 (1980).

Hollerbach, J. M., The Minimum Energy Movement for a Spring Muscle Model, *Artificial Intelligence Laboratory, Massachusetts Institute of Technology*, 424 (1977).

Hollerbach, J.M., A Recursive Formulation of Lagrangian Manipulator Dynamics, *I.E.E.E. Trans. Systems, Man, Cybernetics*, **10**, 11 (1980).

Hollerbach, J.M. Dynamic Interations Between Limb Segments During Planar Arm Movement, *Biol. Cybernetics* **44** (1982).

Hollerbach, J.M., Wrist-Partitioned Inverse Kinematic Accelerations and Manipulator Dynamics, *Int. J. Robotics Res.*, **2**, 4 (1983).

Hollerbach, J.M., Dynamic Scaling of Manipulator Trajectories, *A.S.M.E. J. Dyn. Syst. Meas. Contr.*, **106**, 1 (1984-a).

Hollerbach, J.M., Optimal Kinematic Design for a Seven Degree of Freedom Manipulator, *Int. Symp. of Robotics Research*, Kyoto (1984-b).

Hollis, R.L., A Planar XY Robotic Fine Positioning Device, *I.E.E.E. Int. Conf. Robotics and Automation*, St. Louis (1985).

Holzmann, W. and McCarthy, J. M., Computing the Friction Forces Associated with a Three Fingered Grasp, *I.E.E.E. Int. Conf. Robotics and Automation*, St. Louis (1985).

Hooker, W. W., A Set of R Dynamical Attitude Equations for An Arbitrary N-body Satellite Having R Rotational Degrees of Freedom, *American Institute of Aeronautics and Astronautics Journal*, **8**, 7 (1970).

Hooker, W. W. and Marguiles, G., The Dynamical Attitude Equations for an N-body

Satellite, *J. Astronautical Sciences*, **12**, 4 (1965).

Horn, B.K.P., Robot Vision, *McGraw Hill* (1986).

Horn, B. K. P., Kinematics, Statics, and Dynamics of Two-D Manipulators, *M.I.T. Artificial Intelligence Laboratory*, 99 (1975).

Horn, B.K.P., Hirokawa, P. K. and Varizani, V. V., Dynamics of a Three Degree-of-Freedom Kinematic Chain, *M.I.T. Artificial Intelligence Laboratory*, 478 (1977).

Hughes, P. C., Dynamics of A Flexible Manipulator Arm for the Space Shuttle, *Amer. Astronautical Soc. and Amer. Inst. of Aeronautics and Astrodynamics Specialist Conf.*, Jackson Hole, Wyoming (1977).

Hughes, P.C., Dynamics of A Chain of Flexible Bodies, *J. Astronautical Sciences*, **27**, 4 (1979).

Huston, R. L., Passerello, C. D. F. and Harlow, M. W. Dynamics of Multirigid-body Systems,*J. Applied Mechanics*, **45** (1978).

Inoue, H. , Computer Controlled Bilateral Manipulator, *Bulletin of the J.S.M.E.*, **14**, 69 (1971).

Inoue, H. Force Feedback In Precise Assembly Tasks, *Artificial Intelligence Laboratory, Massachusetts Institute of Technology*, 308 (1977).

IRIA, Seminar On Languages and Programming Methods for Industrial Robots, *Rocquencourt, France* (1979).

Ishida, T., Force Control In Coordination of Two Arms:, *5th Int. Conf. Artificial Intelligence* (1977).

Isidori, A., De Luca, A. and Nicolo, F., Control of Robot Arms with Elastic Joints via Nonlinear Dynamic Feedback, *I.E.E.E. Conf. Decision and Control*, Fort Lauderdale (1985).

Jakubik, P. and Marton, J., Improved Gross-motion Control for Robot Arms, *8th Triennial I.F.A.C. World Congress*, Kyoto, Japan (1981).

Johnson, T. L., On Feedback Laws for Robotic Systems, *8th I.F.A.C. World Congress*, Kyoto, Japan (1981).

Kane, T. R., and Levinson, D. A., The Use of Kane's Dynamic Equations in Robotics, *Int. J. Robotics Research*, **2**, 3, (1983).

Kahn, M. E., The Near-Minimum-Time Control of Open-loop Articulated Kinematic Chains, *Artificial Intelligence Laboratory, Stanford University*, 106 (1969).

Kahn, M. E. and Roth, B., The Near Minimum-Time Control of Open-loop Articulated

Kinematic Chains, *J. Dynamic Systems, Measurement, Control*, **93** (1971).

Kanade, T., Khosla, P. K. and Tanaka, N. Real-Time Control of CMU Direct-Drive Arm II Using Customized Inverse Dynamics, *I.E.E.E. Conf. On Decision and Control*, Las Vegas (1984).

Kane, T. R., Dynamics of Nonholonomic Systems, *J. Applied Mechanics*, **28** (1961).

Karelin, M. M. and Girel, A. M., Accurate Alignment of Parts for Automatic Assembly, *Russian Engineering Journal*, **47**, 9 (1967).

Kazerooni, H., A Robust Design Method for Impedance Control of Constrained Dynamic Systems, *Doctoral Dissertation, Massachusetts Institute of Technology* (1985).

Khatib, O., Commande Dynamique dans L'Espace Operationnel des Robots Manipulateurs en Presence D'obstacles, *Docteur Ingenieur Thesis, Ecole Nationale Superieure de L'Aeronautique et de L'Espace*, Toulouse, France (1980).

Khatib, O., Dynamic Control of Manipulators in Operational Space, *Sixth IFTOMM Congress on Theory of Machines and Mechanisms, New-Delhi* (1983).

Khatib, O., The Operational Space Formulation and the Analysis, Design, and Control of Robot Manipulators, *3rd Int. Symp. Robotics Res., Gouvieux* (1985).

Khatib, O. and Burdick, J., Dynamic Optimization in Manipulator Design: The Operation Space Formulation, *A.S.M.E. Winter Annual Meeting, Symp. Robotics and Manufacturing Automation*, Miami (1985).

Khorasani, K., Spong, M.W., Invariant Manifolds and Their Application to Robot Manipulators With Flexible Joints. *I.E.E.E. Int. Conf. Robotics and Automation*, St. Louis (1985).

Khosla, P.K., and Kanade, T., An Algorithm to Determine the Identifiable Parameters in the Dynamic Robot Model, *I.E.E.E. Int. Conf. Robotics and Automation*, San Francisco (1986).

Khosla, P.K. and Kanade, T., Parameter Identification of Robot Dynamics, *I.E.E.E. Conf. Decision and Control*, Fort Lauderdale (1985).

Kim, B.K. and Shin, K. G., Minimum-Time Path Planning for Robot Arms and Their Dynamics, *I.E.E.E. Trans. Systems Man and Cybernetics*, **15**, 2 (1985).

Kobayashi, H., On Articulate Hands, *Int. Symp. of Robotics Research*, Kyoto (1984).

Koditschek, D., Natural Control of Robot Arms, *Center for Systems Science*, Yale University (1985).

Kohli, D. and Spanos, J., Workspace Analysis of Regional Structures of Manipulators,

A.S.M.E. J. Mech., Trans., & Auto. in Design, **107**, 2 (1985).

Koivo, A.J. and Guo, T.K., Control of Robotic Manipulator With Adaptive Control, *I.E.E.E. Conf. Decision and Control* (1981).

Kondoleon, A. S., Application of Technology-Economic Model of Assembly Techniques to Programmable Assembly Machine Configuration, *S.M. Thesis, Department of Mechanical Engineering, M.I.T.* (1976).

Konstantinov, M.S., Patarinski, S.P., Zamanov, V.B., and Nenchev, D.N., A Contribution to the Inverse Kinematic Problem for Industrial Robots, *12th Int. Symp. Industrial Robots*, Paris (1982).

Kumar, A., And Waldron, K., The Workspace of a Mechanical Manipulator, *A.S.M.E. J. Mechanical Design*, **103**, 3 (1981).

Kuntze, H. B. and Schill, W.,Methods for Collision Avoidance In Computer Controlled Industrial Robotics, *12th Int. Symp. Industrial Robots*, Paris, France (1982).

Laktionev, N. M. and Andreev, G. Y., Automatic Assembly of Parts, *Russian Engineering Journal*, **46**, 8 (1966).

Landau, I. D., Adaptive Control Techniques for Robot Manipulators — The Status of the Art, *I.F.A.C. Symp. Robot Control*, Barcelona (1985).

Larson, R. C. and Li, V. O. K. Finding Minimum Rectilinear Distance Paths in the Presence of Obstacles, *Operations Research Center, Massachusetts Institute of Technology*, 088-79 (1979).

Lathrop, R.M., Parallelism In Manipulator Dynamics, *I.E.E.E. Conf. On Robotics and Automation*, St. Louis (1985).

Lawrence, P.D. and Lin, W. C., Statistical Decision Making in the Real-time Control of an Arm Aid for the Disabled, *I.E.E.E. Trans. Systems, Man, Cybernetics*, **2**, 1 (1972).

Lee, C.S.G., Robot Arm Kinematics, Dynamics, and Control, *Computer*, **15**, 12 (1982).

Leininger, G.G., and Wang, S., Pole Placement Self-Timing Control of Manipulators, *I.F.A.C. Symp. Computer-Aided Design of Multivariable Technological Systems* (1982).

Lesser, E. and Turai, T., Demonstration of Resolved Motion Rate Control, *S.M. Thesis, Department of Mechanical Engineering, Massachusetts Institute of Technology* (1970).

Lewis, R. A. Autonomous Manipulation On A Robot: Summary of Manipulator Software Functions, *Jet Propulsion Laboratory, California Inst. of Technology*, **33**, 679 (1974).

Lewis, R. A. and Bejczy, A. K., Planning Considerations for A Roving Robot With Arm,

Int. Joint Conf. Artificial Intelligence, Stanford University (1973).

Lieberman, L. I. and Wesley, M. A., AUTOPASS: An Automatic Programming System for Computer Controlled Mechanical Assembly, *I.B.M. J. Research Development*, **21**, 4 (1977).

Liegois, A., Khalil, W., Dumas, J. M. and Renaud, M., Mathematical and Computer Models of Interconnected Mechanical Systems, *Symp. Theory and Practice of Robots and Manipulators*, Warsaw, Poland (1976).

Liegois, A., Fournier, A. and Aldon, M. J. Model Reference Control of High Velocity Industrial Robots, *1980 Joint Automatic Control Conf.*, San Francisco (1980).

Litvin, F. L. and Castelli, V. P., Configurations of Robot's Manipulators and Their Identification, and the Execution of Prescribed Trajectories, Part 1: Basic Concepts, *A.S.M.E. J. Mech., Trans., & Auto. in Design*, **107**, 2 (1985).

Lozano-Perez, T., The Design of A Mechanical Assembly System *Artificial Intelligence Laboratory, Massachusetts Institute of Technology*, 397 (1976).

Lozano-Perez, T., Automatic Planning of Manipulator Transfer Movements, *I.E.E.E. Trans. Systems Man. Cybernetics*, **11**, 10 (1981).

Lozano-Perez, T. Spatial Planning: A Configuration Space Approach, *I.E.E.E. Trans. Computers* (1983).

Lozano-Perez, T., Robot Programming, *I.E.E.E. Proceedings* (1983).

Lozano-Perez, T. Mason, M. T. and Taylor R. M., Automatic Synthesis of Fine-Motion Strategies for Robots, *Int. J. Robotics Research* **3**, 1 (1984).

Lozano-Perez, T. and Wesley, M. A. An Algorithm for Planning Collision-free Paths Among Polyhedral Obstacles, *Comm. ACM*, **22**, 10 (1979).

Lozano-Perez, T. and Winston, P. H., LAMA: A Language for Automatic Mechanical Assembly *5th Int. Joint Conf. Artificial Intelligence, M.I.T.*, Cambridge, MA (1977).

Luenberger. D.G., Optimization by Vector Space Methods, *J. Wiley & Sons*, New York (1969).

Luenberger, D.G., Introduction to Dynamic Systems, *J. Wiley & Sons*, New York (1979).

Luh, J. Y. S., Fisher, W. D. and Paul, R. P. C., Joint Torque Control by A Direct Feedback for Industrial Robots, *20th I.E.E.E. Conf. on Decision and Control*, San Diego (1981).

Luh, J.Y.S. and Gu, Y. L., Lagrangian Formulation of Robotic Dynamics with Dual-Number Transformation for Computational Simplification, *18th Conf. On Info Sciences and Systems*, Princeton (1984).

Luh, J. Y. S. and Lin, C. S., Automatic Generation of Dynamic Equations for Mechanical Manipulators, *1981 Joint Automatic Control Conf.* Charlottesville, VA (1981)

Luh, J. Y. S. and Walker, M. W. Minimum-time Along the Path for a Mechanical Arm, *1977 I.E.E.E. Conf. Decision and Control*, **1**, New Orleans (1977).

Luh, J.Y.S., Walker, M. W. and Paul, R. P. C., On-Line Computational Scheme for Mechanical Manipulators *A.S.M.E. J. Dyn. Syst. Meas. Contr.*, **102** (1980-a).

Luh, J.Y.S., Walker, M. H. and Paul, R. P. C., Resolved Acceleration Control of Mechanical Manipulators, *I.E.E.E. Trans. Automatic Control*, **25**, 3 (1980-b).

Luh, J.Y.S., and Zheng, Y. F., Computation of Input Generalized Forces for Robots with Closed Kinematic Chain Mechanisms, *I.E.E.E. J. of Robotics and Automation*, **1**, 2 (1985).

Lundstrom, G., Gripdon for Industrirobtar Och Hanteringsutrustningar, *Flygtekniska Forsoksanstalten*, AU-1143, (1977).

Lynch, P.M., A Mathematical Model of Mechanical Manipulator Dynamics, *S.M. Thesis, Department of Mechanical Engineering, M.I.T.* (1970).

Maciejewski, A. A. and Klein, C. A., Obstacle Avoidance for Kinematically Redundant Manipulators in Dynamically Varying Environments, *Int. J. Robotics Research*, **4**, 3 (1985).

Marino R. and Nicosia S., On the Feedback Control of Industrial Robots with Elastic Joints: A Singular Perturbation Approach, *University of Rome*, R-84.01 (1984).

Markiewicz, B.R., Analysis of the Computer Torque Drive Method and Comparison with Conventional Position Serve for a Computer-controlled Manipulator, *Jet Propulsion Laboratory, Calif. Inst. of Tech.* TM **33**, 601 (1973).

Markowsky, G. and Wesley, M. A., Fleshing Out Wire Frames, *IBM J. Research and Development*, **24**, 5 (1980).

Markus, L. and Lee, E. B., On The Existence of Optimal Controls, *Trans. A.S.M.E. J. Basic Engineering* (1962).

Marr, D. and Nishihara, H. K., Representation and Recognition of the Spatial Organization of Three Dimensional Shapes, *Artificial Intelligence Laboratory, Massachusetts Institute of Technology*, 416 (1977).

Mason, M.T., Compliance and Force Control of Computer Controlled Manipulators, *Artificial Intelligence Laboratory, Massachusetts Institute of Technology*, 515 (1979).

Mason, M. T. Compliance and Force Control for Computer Controlled Manipulators, *I.E.E.E. Trans. Systems, Man and Cybernetics*, **11**, 6 (1981).

Mason, M.T. Manipulator Grasping and Pushing Operations, *Doctoral Dissertation, Massachusetts Institute of Technology*, (1982).

Mason, M.T. and Salisbury, J.K., Robot Hands and the Mechanics of Manipulation, *M.I.T. Press* (1985).

Mason, P., Dynamic Stiffness and Crossbridge Action in Muscle, *Biophysics of Structure and Mechanism 4* (1978).

Mathur, D., The Grasp Planner, *Dept. of Artificial Intelligence*, University of Edinburgh (1974).

McCallion, H. and Wong, P. C., Some Thoughts on the Automatic Assembly of a Peg and a Hole, *Industrial Robot*, **2**, 4 (1975).

Miller, R. L.,"Local Sensing" , Robotics and Artificial Vision Systems Handbook (Brady & Villers, Eds.), *Benjamin/Cummings* (1986).

Moe, M.L. and Schwartz, J. T., A Coordinated, Proportional Motion Controller for an Upper-extremity Orthotic Device, *3rd Int. Symp. External Control of Human Extremities*, Dubrovnik (1969).

Mohamed, M. G. and Duffy, J., A Direct Determination of the Instantaneous Kinematics of Fully Parallel Robot Manipulators, *A.S.M.E. J. Mech., Trans., & Auto. in Design*, **107**, 2 (1985).

Monster, A.W., On Some Control Problems in the Case of Research Arm-Aid, *Engineering Design Center, Case Western Reserve University*, 7-66-14 (1966).

Moravec, H. P., Visual Mapping by a Robot Rover, *6th Int. Joint Conf. Artificial Intelligence*, Tokyo, Japan (1979).

Moravec, H. P., Obstacle Avoidance and Navigation in the Real World by a Seeing Robot Rover, *Doctoral Dissertation, Stanford University* (1980).

Nakamura, Y., and Hanafusa, H., Task Priority Based Redundancy Control of Robot Manipulators, *Int. Symp. of Robotics Research*, Kyoto (1984).

Nakano, E., Ozaki, S., Ishida, T. and Kato, I., Cooperational Control of the Anthropomorphous Manipulator MELARM, *4th Int. Symp. Industrial Robots*, Tokyo (1974).

Nelson, W. L., Mitra, D. and Boie, R. A., End-Point Sensing and Adaptive Control of a Flexible Robot Arm, *I.E.E.E. Conf. Decision and Control*, Fort Lauderdale (1985).

Nevins, J. L. and Whitney, D. E., The Force Vector Assembler Concept, *1st IFTOMM Symp. Theory and Practice of Robots and Manipulators* (1974-a).

250   Bibliography

Nevins, J. L. and Whitney, D. E., Adaptable-programmable Assembly Systems: An Information and Control Problem, *5th Int. Symp. Industrial Robots*, Chicago (1974-b).

Nevins, J. L. and Whitney, D. E. Computer Controlled Assembly, *Scientific American*, **238**, 2 (1978).

Nevins, J. L., Whitney, D.E., et al., Exploratory Research in Industrial Modular Assembly, *C.S. Draper Laboratory*, Cambridge, MA, R-800 (1974-a).

---. R-850 (1974-b).

---. R-921 (1975).

---. R-966 (1976).

---. R-1111 (1977).

Nevins, J. L., Whitney, D. E. and Simunovic, S. N., System Architecture for Assemby Machines, *C.S Draper Laboratory*, Cambridge, MA, R-764 (1973).

Nevins, J. L., Whitney, D. E. and Woodin, A. E., A Scientific Approach to the Design of Computer Controlled Manipulators, *C.S. Draper Laboratory, Cambridge, MA*, R-837 (1974).

Newman, C. P. and Khosla, P. K., Identification of Robot Dynamics: An Application of Recursive Estimation, *4th Yale Workshop on Applications of Adaptive Systems Theory*, Yale University (1985)

Newman, W. M. and Sproull, R. F. , Principles of Interactive Computer Graphics, *McGraw-Hill*, New York (1973).

Nguyen, P.K. and Ravindran, R., Kinematics and Control of the Space Shuttle Remote Manipulator System, *7th Canadian Congress on Applied Mechanics*, Vancouver, Canada (1977).

Nicosia, S., Nicolo, F. and Lentinit, D., Dynamical Control of Industrial Robots With Elastic and Dissipative Joints, *8th Triennial I.F.A.C. World Congress*, Kyoto, Japan (1981).

Nilsson, N., A Mobile Automaton: An Application of Artificial Intelligence Techniques, *Int. Joint Conf. Artificial Intelligence* (1969).

Nilsson, N., Principles of Artificial Intelligence, *Tioga Publishing*, California (1980).

Nobel, B., Applied Linear Algebra, *Prentice-Hall, Englewood Cliffs, NJ* (1969).

Nof, S.Y.(Editor), Handbook of Industrial Robotics, *John Wiley & Sons* (1985).

Ohwovoriole, M.S. An Extension of Screw Theory and its Application to the Automation of Industrial Assemblies, *Doctoral Dissertation, Stanford University* (1980).

Ohwovoriole, M.S. and Roth, B., A Theory of Parts Mating for Assembly Automation, *Ro. Man. Sy. -81*, Warsaw, Poland (1981).

Okada, T., Computer Control of Multi-Jointed Finger System, *6th Int. Joint Conf. on Artificial Intelligence*, Tokyo (1979).

Orin, D. E., McGhee, R. B., Vukobratovic, M. and Hafloch, G., Kinematic and Kinetic Analysis of Open-chain Linkages Utilizing Newton-Euler Methods, *Mathematical Biosciences* **43**, 1/2 (1979).

Orin, D.E. and Oh, S. Y., Control of Force Distribution in Robotic Mechanisms Containing Closed Kinematic Chains, *A.S.M.E. J. Dynamic Systems, Measurement, Control*, 102 (1981).

Orin, D.E. and Schrader, W.W., Efficient Computation of the Jacobian for Robot Manipulators, *Int. J. Robotics Research* (1984).

Pakie, M., The Belgrade Hand Prosthesis, *Inst. Mechanical Engineers* **183**, 3F (1969).

Park, W.T., Minicomputer Software Organization for Control of Industrial Robots, *1977 Joint Automatic Control Conf.*, San Francisco (1977).

Patwarkhan, A. G., and Soni, A. H., Motion Simulation of an Articulated Robotic Arm Subjected to Static Forces, *A.S.M.E. J. Mechanical Design*, **104**, 2 (1982).

Paul, B., Kinematics and Dynamics of Planar Machinery, *Prentice- Hall*, Englewood Cliffs, NJ (1979).

Paul, R.P., Modelling, Trajectory Calculation, and Serving of A Computer Controlled Arm, *Stanford University, Artificial Intelligence Laboratory*, 177 (1972).

Paul, R.P., Manipulator Path Control, *I.E.E.E. Int. Conf. Cybernetics and Society*, New York (1975).

Paul, R.P., WAVE: A Model-Based Language for Manipulator Control, *Industrial Robot* (1977).

Paul, R.P. Manipulator Cartesian Path Control, *I.E.E.E. Trans. Systems, Man, Cybernetics*, **9**, 702 (1979).

Paul, R.P., Robot Manipulators: Mathematics Programming and Control, *M.I.T. Press*, Cambridge, MA (1981).

Paul, R.P. and Shimano, B., Compliance and Control, *1976 Joint Automatic Control Conf.*, San Francisco (1976).

Paul, R. P. and Stevenson, C. N., Kinematics of Robot Wrists, *Int. J. Robotics Research*, **2**, 1 (1983).

Paul, R.P., Luh, J. Y. S., et al., Advanced Industrial Robot Control Systems, *Purdue University*, RR78-25 (1978).

Pennock, G. R. and Yang, A. T., Application of Dual-Number Matrices to the Inverse Kinematics Problem of Robot Manipulators, *A.S.M.E. J. Mech., Trans., & Auto. in Design*, **107**, 2 (1985-a).

Pennock, G. R. and Yang, A. T., Instantaneous Kinematics of Three-Parameter Motions, *A.S.M.E. J. Mech., Trans., & Auto. in Design*, **107**, 2 (1985-b).

Pfister, G., On Solving the FINDSPACE Problem Or How to Find Where Things Aren't, *Artificial Intelligence Laboratory, Massachusetts Institute of Technology*, 113 (1973).

Pieper, D. L., The Kinematics of Manipulators Under Computer Control, *Doctoral Dissertation, Stanford University* (1968).

Pieper, D. L. and Roth, B., The Kinematics of Manipulators Under Computer Control, *2nd Int. Conf. Theory of Machines and Mechanisms, Warsaw* (1969).

Polit, A. and Bizzi, E., Characteristics of Motor Programs Underlying Arm Movements in Monkeys, *J. Neurophysilogy*, **42**, 1 (1979).

Popov, V.M., Absolute Stability of Nonlinear Control Systems of Automatic Control, *Automation and Remote Control*, **22** (1962).

Popplestone, R.J., Specifying Manipulation in Terms of Spatial Relationships, *Department of Artificial Intelligence, University of Edinburgh*, 117 (1979).

Popplestone, R.J., Ambler, A. P. and Bellos, I., RAPT, A Language for Describing Assemblies, *Industrial Robot*, **5**, 3 (1978).

Popplestone, R.J., Ambler, A. P. and Bellos, I., An Interpreter for a Language for Describing Assemblies, *Artificial Intelligence*, **14**, 1 (1980).

Preparata, F. and Hong, S., Convex Hulls of Finite Sets of Point in Two and Three Dimensions, *Comm. ACM*, **20**, 2 (1977).

Prigogine, I., Introduction to Thermodynamics of Irreversible Processes, 3rd Ed., *Wiley-InterScience, New York* (1967).

Purbrick, J.A., A Force Transducer Employing Conductive Silicone Rubber, *1st Int. Conf. Robot Vision and Sensory Controls* (1981).

Purbrick, J.A., A Multi-Axis Force Sensing Finger, *2nd Ann. A.S.M.E. Conf. Computers* (1982).

Raibert, M.H., Analytical Equations Vs. Table Look-up for Manipulation: A Unifying Concept,*I.E.E.E. Conf. Decision and Control*, New Orleans (1977).

Raibert, M. H., A Model for Sensory Motor Control and Learning, *Biological Cybernetics,***29** (1978).

Raibert, M. H., Legged Robots that Balance, *M.I.T. Press* (1986).

Raibert, M. H. and Craig J. J., Hybrid Position/force Control of Manipulators, *J. Dynamic Systems, Measurement, Control 102* (1981).

Raibert, M. H. and Horn, B. K. P., Manipulator Control Using the Configuration Space Method, *Industrial Robot*, 5, 2 (1978).

Raibert, M.H. and Tanner, J. F., Design and Implementation of a VLSI Tactile Sensing Computer, *Int. J. Robotics Research*, 1, 3 (1982).

Reddy, D. R. and Rubin, S., Representation of Three-dimensional Objects, *Department of Computer Science, Carnegie-Mellon University*, 78-113 (1978).

Renaud, M., An Efficient Iterative Analytical Procedure for Obtaining a Robot Manipulator Dynamic Model, *Int. Symp. Robot. Res., Bretton Woods* (1983).

Requicha, A.A.G., Representation of Rigid Solids: Theory, Methods, and Systems, *Computing Surveys*, 12, 4 (1980).

Roberts, L.G. Machine Perception of Three-dimensional Solids, *Optical and Electro-Optical Information Processing* (1962).

Rooney, J., A Survey of Representations of Spatial Rotation About a Fixed Point, *Environment and Planning B 4* (1978a).

Rooney, J., On the Three Types of Complex Number and Planar Transformations, *Environment and Planning B 5* (1978b).

Rosen, C., et al., Exploratory Research in Advanced Automation, Report to The National Science Foundation , *Stanford Research Institute*, Menlo Park, CA (1976a).

Rosen, C., et al., Machine Intelligence Research Applied to Industrial Automation, Report of The National Science Foundation, *Stanford Research Institute*, Menlo Park, CA (1976b).

Rosen, C., et al., Machine Intelligence Research Applied to Industrial Automation, 6th Report of the National Science Foundation, Grant APR75-10374, *Stanford Research Institute*, Menlo Park, CA (1976c).

Roth, B., Performance Evaluation of Manipulators from a Kinematic Viewpoint, *NBS Special Publication: Performance Evaluation of Programmable Robots and Manipulators* (1975).

Roth, B., Screws and Wrenches That Cannot Be Bought at a Hardware Store, *Int. Symp. of Robotics Research*, Bretton Woods (1983).

Sahar and Hollerbach, Planning of Minimum-Time Trajectories for Robot Arms, *Artificial Intelligence Laboratory, Massachusetts Institute of Technology*, 804 (1984).

Salisbury, J.K. Active Stiffness Control of a Manipulator in Cartesian Coordinates, *I.E.E.E.*

Conf. Decision and Control, Albuquerque, New Mexico (1980).

Salisbury, J.K., Kinematic and Force Analysis of Articulated Hands, *Doctoral Dissertation, Stanford University* (1982).

Salisbury, J.K., Design and Control of an Articulated Hand, *Int. Symp. On Design and Systems*, Tokyo (1984).

Salisbury, J.K., and Craig, J. J., Articulated Hands: Force Control and Kinematic Issues, *Int. J. Robotics Research*, **1**, 1 (1982).

Salmon, M., Robot Technology at Olivetti: The Sigma System, *Olivetti*, Milan (1976).

Samet, H., Region Representation: Quadtrees from Boundary Codes, *Comm. ACM*, **23**, 3 (1980).

Saridis, G.N., Intelligent Robotic Control, *1981 Joint Automatic Control Conf.*, Charlottesville, VA (1981).

Saridis, G.N. and Stephano, H. E., Hierarchically Intelligent Control of a Bionic Arm, *I.E.E.E. Conf. Decision and Control* (1975).

Savischenko, V. M. and Bespalov, V. G., Orientation of Components for Automatic Assembly, *Russian Engineering Journal*, **45**, 5 (1965).

Scheinman V. C., Design of a Computer Controlled Manipulator, *Stanford Artificial Intelligence Laboratory*, 92 (1969).

Schmitt, D., Soni, A. H., Srinivasan, V., and Naganathan, G., Optimal Motion Programming of Robot Manipulators, *A.S.M.E. J. Mech., Trans., & Auto. in Design*, **107**, 2 (1985).

Schwartz, J.T. and Sharir, M., On the Piano Movers Problem 1: The Case of a Two-Dimensional Rigid Polygonal Body Moving Amidst Polygonal Barriers, *Department of Computer Science, Courant Institute of Mathematical Sciences, New York University*, 39 (1981).

Schwartz, J.T. and Sharir, M., On the Piano Movers Problem 1: General Properties for Computing Topological Properties of Real Algebraic Manifolds, *Department of Computer Science, Courant Institute of Mathematical Sciences, New York University*, 41 (1982).

Seltzer, D.S., Use of Sensory Information for Improved Robot Learning, *Autofact*, Detroit (1979).

Serra, J., Image Analysis and Mathematical Morphology, *Academic Press* (1982).

Shiller, Z. and Dubowsky, S., On the Optimal Control of Robotic Manipulators With Actuator and End-Effector Constraints, *I.E.E.E. Int. Conf. On Robotics and Automation* (1985).

Shimano, B., "Improvements in the AL Runtime System", *Exploratory Study of Computer Integrated Assembly Systems, Artificial Intelligence Laboratory, Stanford University*, (1976).

Shimano, B., "Force Control", *Exploratory Study of Computer Integrated Assembly Systems, Artificial Intelligence Laboratory, Stanford University* (1977).

Shimano, B., The Kinematic Design and Force Control of Computer Controlled Manipulators, *Artificial Intelligence Laboratory, Stanford University* (1978).

Siljak, D.D., Large Scale Dynamic Systems, *North-Holland* (1978).

Silver, D. B., The Little Robot System, *Artificial Intelligence Laboratory, Massachusetts Institute of Technology*, 273 (1973).

Silver, D. B. On the Equivalence of Lagrangian and Newton-Euler Dynamics for Manipulators, *Int. J. Robotics Research*, **1**, 2 (1982).

Simon, H. A., The Sciences of the Artificial, 2nd Ed., *M.I.T. Press* (1981).

Simpson, D.C. An Experimental Design for a Powered Arm Prosthesis, *Health Bulletin 23* (1965).

Simunovic, S.N., Force Information in Assembly Process, *5th Int. Symp. Industrial Robots*, Chicago (1975).

Simunovic, S.N., An Information Approach to Parts Mating,*Doctoral Dissertation, Department of Electrical Engineering, Massachusetts Institute of Technology* (1979).

Singh, S. N. and W. J. Rugh, Decoupling in a Class of Nonlinear Systems by State Variable Feedback, *J. Dynamic Systems, Measurement, Control* (1972).

Skinner, F., Multiple Prehension Hand For Assembly Robots, *5th Int. Symp. Industrial Robots*, Chicago (1975).

Slotine, J.-J. E., Tracking Control of Nonlinear Systems using Sliding Surfaces, *Doctoral Dissertation, Massachusetts Institute of Technology* (1983).

Slotine, J.-J.E., Suction Control of Robot Manipulators, *18th Conf. Info. Sciences and Systems*, Princeton, NJ (1984a).

Slotine, J.-J. E. Robustness Issues in the Control of High-Performance Robots, *I.E.E.E. Conf. Decision and Control*, Las Vegas (1984b)

Slotine, J.-J. E. Sliding Controller Design for Nonlinear Systems, *Int. J. Control.*, **40**, 2 (1984c).

Slotine, J.-J. E. Robustness Issues in Robot Control, *I.E.E.E. Int. Conf. Robotics and Automation*, St. Louis (1985a).

Slotine, J.-J. E. The Robust Control of Robot Manipulators, *Int. J. Robotics Research*, **4**, 2 (1985b).

Slotine, J.-J. E. and Coetsee, J. A., Adaptive Sliding Controller Synthesis for Nonlinear Systems, *Int. J. Control* (1986).

Slotine, J.-J. E. and Sastry S. S., Tracking Control of Nonlinear Systems Using Sliding Surfaces With Applications to Robot Manipulators, *Int. J. Control*, **39**, 2 (1983).

Slotine, J.-J.E. and Spong, M.W., Robust Robot Control with Bounded Input Torques, *Int. J. Robotics Systems*, **2**, 4 (1985).

Slotine, J.-J. E., and Yoerger, D.R., Inverse Kinematics Algorithm for Redundant Manipulators, *Int. J. Robotics and Automation*, **1**, 2 (1986).

Spanos, J. and Kohli, D., Workspace Analysis of Regional Structures of Manipulators, *A.S.M.E. J. Mech., Trans., & Auto. in Design*, **107**, 2 (1985).

Spong, M.W., Thorp, J.S., and Kleinwaks, J.M., The Control of Robot Manipulators with Bounded Input, *School of Electrical Engineering, Cornell University* (1984).

Spong, M.W., and Vidyasagar M., Robust Tracking and Disturbance Rejection for Robot Manipulators, *I.E.E.E. Conf. Decision and Control*, Fort Lauderdale (1985).

Stepanenko, Y. and M. Vukobratovic, Dynamics of Articulated Open-chained Active Mechanisms, *Mathematical Biosciences*, **28** (1976).

Stoyan, Y.G. and Ponomarenko, L. D., A Rational Arrangement of Geometric Bodies in Automated Design Problems, *Engineering Cybernetics*, **16**, 1 (1978).

Sugimoto, K., Determination of Joint Velocities of Robots by Using Screws, *A.S.M.E. J.. of Mech., Trans., Auto. in Design*, **106**, 2 (1984).

Sugimoto, K., and Duffy, J, Determination of Extreme Distances of a Robot Hand, Robot Arms with Special Geometry, *A.S.M.E. J.. Mechanical Design*, **103**, 4 (1981).

Sweet, L. M. and Good, M. C., Redefinition of the Robot Control Problem: Effects of Plant Dynamics, Drive System Constraints, and User Requirements, *I.E.E.E. Conf. Decision and Control*, Las Vegas (1984).

Symon, K.R. Mechanics, *Addison Wesley* Reading, MA (1971).

Takase, K., Inoue, H., Sato, K. and Hagiwara, S., The Design of an Articulated Manipulator with Torque Control Ability, *4th Int. Symp. Industrial Robots, Tokyo* (1974).

Takegaki, M. and Arimoto, S., A New Feedback Method for Dynamic Control of Manipulators, *J. Dynamic Systems, Measurement, Control*, **102** (1981).

Takeyasu, K., Goto, T. and Inoyama, T., Precision Insertion Control Robot its Application, *J. Engineering for Industry B98-4* (1976).

Taylor, R.H., The Synthesis of Manipulator Control Programs from Task-level Specificiations, *Artificial Intelligence Laboratory, Stanford University*, 282 (1976).

Taylor, R.H. Planning and Execution of Straight-line Manipulaor Trajectories, *IBM Thomas J. Watson Research Center, Yorktown Heights*, New York, Research Report RC6657 (1977).

Taylor, R.H., Planning and Execution of Straight-line Manipulator Trajectories, *IBM J. Research and Development*, **23** (1979).

Thomas, M., and Tesar, D., Dynamic Modeling of Serial Manipulator Arms, *A.S.M.E. J. of Dyn. Sys. Meas. Control*, **104**, 3 (1982).

Thomas, M., Yuan-Chou, H. C., and Tesar, D., Optimal Actuator Sizing for Robotic Manipulators Based on Local Dynamic Criteria, *A.S.M.E. J. Mech., Trans., Auto. in Design*, **107**, 2 (1985).

Thompson, A.M. The Navigation System of the JPL Robot, *5th Int. Joint Conf. Artificial Intelligence, Massachusetts Institute Of Technology* (1977).

Townsend, A.L., Linear Control Theory Applied to a Mechanical Manipulator, *SB Thesis, Department of Mechanical Engineering, Massachusetts Institute of Technology* (1972).

Truckenbrodt, A., Truncation Problems in the Dynamics and Control of Flexible Mechanical Systems, *8th Triennial I.F.A.C. World Congress*, Kyoto, Japan (1981).

Tsai, Y. C., and Morgan, A. P., Solving the Kinematics of the Most General Six- and Five-Degree-of-Freedom Manipulators by Continuation Methods, *A.S.M.E. J. Mech., Trans., & Auto. in Design*, **107**, 2 (1985).

Tsai, Y. C., and Soni, A. H., Accessible Region and Synthesis of Robot Arms, *A.S.M.E. J. Mechanical Design*, **103**, 4 (1981).

Udupa, S.M., Collision Detection and Avoidance in Computer Controller Manipulators, *5th Int. Joint Conf. Artificial Intelligence, Massachusetts Institute of Technology*, (1977a).

Udupa, S.M., Collision Detection and Avoidance in Computer Controller Manipulators, *Doctoral Dissertation, Department of Electrical Engineering, California Institute of Technology* (1977b).

Uicker, J.J., On the Dynamic Analysis of Spatial Linkages Using 4 by 4 Matrices, *Doctoral Dissertation, Department of Mechanical Engineering and Astronautical Sciences*, Northwestern University (1965).

Uicker, J.J., Denavit, J. and Hartenbert, R. S., An Iterative Method for the Displacement Analysis of Spatial Mechanisms, *J. Applied Math.*, **31** (1964).

Utkin, V.I, Equations of Sliding Mode in Discontinuous Systems I,II, *Automation and Remote Control* (1972).

Utkin, V.I., Variable Structure Systems With Sliding Mode: A Survey, *I.E.E.E. Trans. Automatic Control AC-22* (1977).

Utkin, V.I., Sliding Modes and Their Application to Variable Structure Systems, *MIR Publishers*, Moscow (1978).

Vidyasagar, M., Nonlinear System Analysis, *Prentice-Hall*, Englewood Cliffs, NJ (1978).

Vukobratovic, M., Dynamics of Active Articulated Mechanisms and Synthesis of Artificial Motion, *Mechanism and Machine Theory*, **13** (1978).

Vukobratovic, M., Stokic, D. and Hristic, D., New Control Concept of Anthropomorphic Manipulators, *Mechanism and Machine Theory*, **12** (1977).

Waldron, K. J., Wang, S.-L., and Bolin, S. J. A Study of the Jacobian Matrix of Serial Manipulators, *A.S.M.E. J. Mech. Trans. & Auto. in Design*, **107**, 2 (1985).

Walker, M.W. and Orin, D. E., Efficient Dynamic Computer Simulation of Robot Mechanisms, *A.S.M.E. J. Dynamic Systems, Measurement, Control*, **104** (1982).

Wang, T. and Kohli, D., Closed and Expanded Form of Manipulator Dynamics Using Lagrangian Approach, *A.S.M.E. J. Mech. Trans. & Auto. in Design*, **107**, 2 (1985).

Wang, S.S.M. and Will, P. M., Sensors for Computer Controlled Mechanical Assembly, *Industrial Robot* (1978).

Warnock, J.E., A Hidden-Surface Algorithm for Computer Generated Half-tone Pictures, *Computer Science Department, University of Utah*, 4-15 (1969).

Waters, R.C., Mechanical Arm Control, *Artificial Intelligence Laboratory, Massachusetts Institute of Technology*, 549 (1979).

Watson, P.C., A Multidimensional System Analysis of the Assembly Process as Performed by a Manipulator, *C. S. Draper Laboratory*, Cambridge, MA (1976).

Watson, P.C., Mechanical Arm Control, *Artificial Intelligence Laboratory, Massachusetts Institute of Technology*, 549 (1979).

Watson, P.C. and Drake, S. H., Pedestal and Wrist Force Sensors for Automatic Assembly, *5th Int. Symp. Industrial Robots*, Chicago, (1975).

Waxman, A.M. and Ullman, S., Surface Structure and Three-Dimensional Motion from

Image Flow Kinematics, *Int. J. of Robotics Research*, **4**, 3 (1985).

Waxman, A.M. and Wohn, K., Contour Evolution, Neighborhood Deformation, and Global Image Flow: Planar Surfaces in Motion, *Int. J. of Robotics Research*, **4**, 3 (1985).

Wesley, M.A., et al., A Gemoetric Modeling System for Automated Mechanical Assembly, *IBM J. Research and Development*, **24** (1980).

West, H., and Asada, H., Kinematic Analysis and Mechanical Advantage of Manipulators Constrained by Contact with the Environment, *A.S.M.E. Winter Annual Meeting*, Miami (1985).

Whitney, D.E., Resolved Motion Rate Control of Manipulators and Human Prostheses, *I.E.E.E. Trans. Man-Machine Systems* **10**, (1969-a)

Whitney, D.E. Optimum Stepsize Control for Newton-Raphson Solution of Nonlinear Vector Equations, *I.E.E.E. Trans. Automatic Control*, **14**, 5 (1969-b).

Whitney, D.E. Use of Resolved Rate to Generate Torque Histories for Arm Control, *C.S. Draper Laboratory*, MAT-54 (1972).

Whitney, D.E. The Mathematics of Coordinated Control of Prostheses and Manipulators, *J. Dynamics Systems, Measurement, Control* (1972).

Whitney, D.E., Force Feedback Control of Manipulator Fine Motions, *1976 Joint Automatic Control Conf.*, San Francisco (1976).

Whitney, D.E. Force Feedback Control of Manipulator Fine Motions, *A.S.M.E. J. Dynamic Systems, Measurement, Control* (1977).

Whitney, D.E., Quasi-Static Assembly of Compliancy Supported Right Parts, *A.S.M.E. J. Dynamic Systems, Measurement, Control*, **104** (1982).

Whitney, D.E., Historical Perspectives and State of the Art in Robot Force Control, *I.E.E.E. Int. Conf. Robotics and Automation, St. Louis* (1985).

Whitney, D.E. and Edsall, A.C., Modeling Robot Contour Processes, *2nd Int. Sym of Robotics Research* (1984).

Whitney, D.E. and Junkel, E. F., Applying Stotchastic Control Theory to Robot Sensing, Teaching, and Long Term Control, *Amer. Control Conf.*, Alexandria, VA (1982).

Wiener, N., Cybernetics, or Control and Communication in the Animal and the Machine, *M.I.T. Press* (1961).

Will, P.M. and Grossman, D. D., An Experimental System for Computer Controlled Mechanical Assembly, *I.E.E.E. Trans. Computers*, **24**, 9 (1975).

Williams, R.J. and Seireg, A., Interactive Modeling and Anlysis of Open or Closed Loop Dynamic Systems with Redundant Actuators, *Mechanical Design*, **101**, 3 (1979).

Wingham, M., Planning How to Grasp Objects in a Cluttered Environment, *M.Ph. Thesis*, Edinburgh (1977).

Winston, P.H., Artificial Intelligence, *Addison-Wesley* (1984).

Wirta, R.W. and Taylor, D. R., Multiple-axis Myoelectrically Controlled Prosthetic Arm, *Krusen Center, Moss Rehabilitation Hospital, Philadelphia, Final Report* (1970).

Wittenburg, J., Dynamics of Systems of Rigid Bodies, *B.G. Teubner*, Stuttgart (1977).

Wolovich, W.A., Linear Multivariable Systems, *Springer-Verlag*, New York (1974).

Wu, Chi-haur, Compliance Control of a Robot Manipulator Based on Joint Torque Servo, *Int. J. Robotics Research*, **4**, 3 (1985).

Yang, D.C.H. and Lai, Z.C., On the Dexterity of Robotic Manipulators-Service Angle, *A.S.M.E. J. Mech. Trans. & Auto. in Design*, **107**, 2 (1985).

Yang, A.T. and Freudenstein, F., Application of Dual-number Quaternion Algebra to the Analysis of Spatial Mechanisms, *J. Applied Mechanics*, **86** (1964).

Yoerger, D.R. and Slotine, J.-J.E., Robust Trajectory Control of Underwater Vehicles, *I.E.E.E. J. Oceanic Eng.*, **10**, 4 (1985).

Yoerger, D.R. and Slotine, J.-J. E., Task-resolved Motion Control of Vehicle Manipulator Systems, *Int. J. Robotics and Automation*, **1**, 2 (1986).

Yoshikawa, T., Dynamic Manipulability of Robot Mechanisms, *I.E.E.E. Int. Conf. Robotics and Automation*, St. Louis (1985).

Youcef-Toumi, K., and Asada, H., The Design of Arm Linkages with Decoupled and Configuration-Invariant Inertia Tensors, *A.S.M.E. Winter Annual Meeting, Symp. Robotics and Manufacturing Automation*, Miami (1985).

Young, K.-K. D., Control and Trajectory Optimization of a Robot Arm, *University of Illinois, Urbana, CSL Report R-701* (1975).

Young, K.-K. D., Control and Optimization of Robot Arm Trajectories, *I.E.E.E. Milwaukee Symp. Automatic Computation and Control*, Milwaukee (1976).

Young, K.-K. D., Controller Design for A Manipulator Using Theory of Variable Structure Systems, *I.E.E.E. Trans. Systems, Man, Cybernetics*, **8**, 2 (1978).

Young, K.-K.D., Kokotovic, P. K. and Utkin, V. I., A Singular Perturbation Analysis of High Gain Feedback Systems, *I.E.E.E. Trans. Automatic Control*, **22** (1977).

Zahalak, G.L. and Heyman, S. J., A Quantitative Evaluation of the Frequency-response Characteristics of Active Human Skeletal Muscle in Vivo, *J. Biomechanical Engineering* (1979).

Zalucky, A. and Hardt, D.E., Active Control of Robot Structure Deflections, *A.S.M.E. J. Dyn. Syst. Meas. Contr.*, **106**,1 (1984).

INDEX